TO FORGIVE DESIGN

To Forgive Design

UNDERSTANDING FAILURE

Henry Petroski

The Belknap Press of Harvard University Press

CAMBRIDGE, MASSACHUSETTS

LONDON, ENGLAND   2012

Library of Congress Cataloging-in-Publication Data

Petroski, Henry.

 To forgive design : understanding failure / Henry
Petroski.

   p.   cm.

 Includes bibliographical references and index.

 ISBN 978-0-674-06584-0 (alk. paper)

 1. System failures (Engineering)   I. Title.

TA169.5.P48   2012

620′.00452—dc23   2011044194

To Catherine, this bookend for an anniversary

CONTENTS

PREFACE

It has been two-and-a-half decades since the publication of my first book, *To Engineer Is Human: The Role of Failure in Successful Design,* and I am pleased that it is still being read and commented upon. I attribute at least part of that book's staying power to its accessible discussion of underlying principles of engineering design and how they can be illustrated through real-world examples of successes and failures, many of which were contemporary at the time of the writing. Although the overarching principles developed in *To Engineer Is Human* remain relevant, its examples are limited largely to mechanical and structural failures. I did not have much to say about the interaction of people and machines or the complexities of systemic failures that are not attributable to design alone.

Since the appearance of *To Engineer Is Human,* notable and highly visible accidents have occurred, including the catastrophic loss of two space shuttles; the rush-hour collapse of the interstate highway bridge in Minneapolis; the embarrassing and tragic developments associated with Boston's epic Big Dig; the destruction of the *Deepwater Horizon* oil rig and subsequent persistent oil leak in the Gulf of Mexico; and the rash of construction crane accidents that indiscriminately claimed the lives of workers and ordinary citizens alike. These cases not only further my earlier observations on the interrelationship between success and failure but also illuminate the additional aspects of the engineering enterprise that are embodied in systems and organizations, which are a focus of the present book.

In this sequel to *To Engineer Is Human,* I look at design from a broader perspective and go beyond that fundamental process in identifying additional causes of failure in made things and systems. I develop a number of fresh case studies of landmark accidents and incidents whose occurrence has had a profound effect on how we deal with technology today. These paradigms of failure analysis provide insights into the complexities of design and its aftermath, which involve matters of use, abuse, and regulation of the artifactual systems that populate our built environment.

Some of the cases I deal with in *To Forgive Design* are historical in nature, but in light of new evidence they have been reinterpreted in recent years to yield fresh insights into the root causes of their failure and thus offer new lessons to be learned. The persistence of engineers in investigating what might have seemed to be closed cases involving obsolete technology highlights the fact that there are timeless aspects of the engineering enterprise that are independent of the technology itself and that can be illuminated by a thoughtful reappraisal of even a centuries-old failure to yield new lessons from old examples.

As they did in *To Engineer Is Human,* bridges and their failures play a prominent role in *To Forgive Design.* This reflects in part my own predilection for these products of pure engineering, albeit ones embedded in social, fiscal, and regulatory systems. I see the stories of bridges, as embodied in their conception, design, use, and sometimes failure, as being at the same time engaging and paradigmatic. Some of the bridges and bridge collapses described in this book may not be as familiar as others, but each of their stories adds to our understanding of the many facets of failure and its consequences. I revisit the infamous Tacoma Narrows Bridge failure because, although the moving images of its final writhing and collapse continue to be widely viewed, the story of its origins and legacy are seldom told in sufficient detail to make clear the root cause of the dramatic collapse.

I have been thinking, writing, and lecturing about failure and its implications for a good deal of my professional career, and I trace my fascination with the subject to conscious and unconscious influences, including some encountered in my childhood, student, and graduate-school days. These experiences are recounted here and there in this book as a means of humanizing the subject and viewing it from a different perspective. Failure analysis can be a cold and myopic pursuit; by looking at it through a wide-angle lens we can better see it in the context of our own lives and the lives of others, which it can affect profoundly.

For some time now, my regular columns in *American Scientist*, the magazine of the scientific and engineering research society Sigma Xi, and in *Prism*, that of the American Society for Engineering Education, have provided me with ongoing forums for exploring case studies and professional experiences relating to failure, among other topics, but space limitations in those venues do not allow for a full development of overarching themes. I am grateful to Michael Fisher for giving me this opportunity to reflect on failure at length in this extended essay. It has been a pleasure to be working once again with Michael and his colleagues at Harvard University Press.

During the time that I was developing some of the ideas underlying this book, I was fortunate one semester at Duke to have had weekly meetings with Michael Schallmo, a civil engineering student who was taking an independent-study course in failure analysis under my direction. Our wide-ranging discussions of the subject, including some of the case studies contained herein, provided me with a valuable resource and sounding board in the months before I sat down to begin writing. I am also grateful to Kenneth Carper, editor-in-chief of the *Journal of Performance of Constructed Facilities*, who read the entire manuscript of this book and provided many thoughtful and helpful comments and suggestions. Whatever errors it might contain are, of course, my responsibility alone.

As I have been with all of my books, I am indebted to my wife, Catherine Petroski, who not only has passed on to me much relevant material from her omnivorous reading but also has once again been generous enough with her own time to give my work a critical reading. Thank you, Catherine.

TO FORGIVE DESIGN

Good nature and good sense must ever join;
To err is human, to forgive, divine.

—Alexander Pope, *An Essay on Criticism*

ONE

By Way of Concrete Examples

When Continental Connection Flight 3407 took off from Newark Liberty Airport for Buffalo Niagara Airport on February 12, 2008, it should have been a short hop in the Bombardier Dash 8 aircraft. However, on that cold winter evening the upstate New York area was experiencing freezing rain, which tends to coat airplane wing surfaces with ice. As it builds up, the ice not only adds weight but also alters the aerodynamic shape of the wing, and this naturally affects the way the plane performs in the air. Because ice buildup can produce dangerous flying conditions, airplane wings are designed with features that can break up the accumulated ice. One such feature heats the wing surface; another, known as a de-icing boot, is incorporated into the structure's leading edge. When activated, the boot expands and thus cracks and sheds any ice that has formed. Unfortunately, like all designs, this scheme has limitations, and ice that is located too far back from the leading edge may not be removed by the action of the expanding boot.[1]

The regional twin-engine turboprop plane never made it to Buffalo that evening; it crashed into a house about five miles short of the runway. All forty-nine people on board were killed, as was a person in the house. Early speculation was that somehow the icing conditions were not adequately dealt with, by the wing systems or by the flight crew. After the crash, the cockpit voice recorder and flight data recorder were retrieved, and they revealed that the pilot and first officer discussed the buildup of ice on the windscreen and wings of the aircraft and had turned on the de-icing equipment. However, shortly afterward the plane showed signs of entering a stall. When this happens, the control stick is supposed to shake to warn the pilot. Since the autopilot was on in the Continental flight, the plane should also have self-adjusted to maintain lift, but the autopilot would have shut itself off if the situation was beyond the limits of its control. In such circumstances, the human pilot is supposed to take over and fly the plane manually. This is all by design. One theory about why the pilot did not take appropriate corrective measures involved the possibility that the control stick did not shake, which in turn pointed to an equipment malfunction or a "design defect." Another early report, which described the appropriate response to a stall as a "stick pusher" action, speculated that the pilot may have overreacted by pulling the nose too far up, a decision that would "raise questions about training and airplane design."[2]

Those speculations were made within a week of the crash. The National Transportation Safety Board (NTSB), which is responsible for the in-depth investigation of accidents, typically takes much longer to reach its ultimate conclusions. In its report on the Buffalo case, released just ten days short of a year after the incident, the NTSB did not describe hardware or software design failures but rather painted of the crew "a picture of complacency resulting in catastrophe." The plane's forty-seven-year-old captain, whom a newspaper report termed "inept," had a history of failed flight tests and "tended to overreact in using a plane's controls." Evidently he—who had begun flying the Dash 8 only two months prior to the accident—and his twenty-four-year-old "first officer who would not challenge him" did not monitor the plane's airspeed on their approach to the Buffalo area, and they allowed the speed to get so low that an alarm was triggered. The control stick did respond by shaking, as it was designed to do, but the pilot pulled back instead of pushing forward on the stick, and his repeated incorrect responses made matters worse. The NTSB found no problems with the plane's engines or systems, and it concluded that the weather was typical for Buffalo at that time of year and that icing conditions were not severe enough to have brought down the plane.[3]

One member of the investigatory board believed that pilot fatigue contributed to the accident. The pilot had spent the night before the fatal flight in the crew lounge; the co-pilot, who had a bad cold, had spent it on redeye flights from Seattle to Newark. However, whether the crew was tired or not, its behavior in the cockpit violated Federal Aviation Administration (FAA) policies. The co-pilot sent text messages, which was against both FAA and airline policy. The pilot reportedly "began and dominated long and irrelevant conversations both at takeoff and on approach to Buffalo," which is also against FAA rules. One board member summed up the situation in the cockpit as one in which the pilot and co-pilot "squandered their time" instead of "managing the plane." The accident did not have to happen; there evidently was no problem with the plane's design.[4]

Tragic accidents and failures always bring renewed focus on the nature and reliability of engineering structures and technological systems of all kinds. This has certainly been true in recent decades. A space shuttle exploded upon launch and another disintegrated upon reentry into the Earth's atmosphere; what were once the tallest buildings in the world collapsed shortly after being struck by aircraft that ignited uncontrollable fires; a hurricane brought a city to its knees, killing over 1,300 people and leaving tens of thousands more homeless and hungry; an earthquake claimed as many as a quarter million lives in one of the poorest countries on Earth;

the world's largest automaker recalled millions of its cars because gas pedals seemed to be sticking in the acceleration mode and brakes did not appear to be operating properly; and an explosion on a drilling rig in the Gulf of Mexico was followed by an undersea oil leak that lasted for months and produced an environmental catastrophe.

The list of accidents, tragedies, disasters, embarrassments, and outright failures is long and varied, and it likely will continue to grow even as this book is being written and read. Indeed, the list is but the latest installment in an ongoing accumulation of marks against engineering and technology. But the list did not begin in the last century, nor will it likely end during the new millennium. Near and outright failures have always been part of the human endeavor known as engineering and its collective achievements known as technology. Moreover, because we are human and by nature fallible, we can count on bad things continuing to occur, often when we least expect them. The best we might hope for is to maximize our ability to prevent them and thereby minimize their future occurrence. These goals can be achieved by understanding not only how and why failures have happened in the past and continue to happen today but also, and perhaps more fundamentally, the nature of failure itself.

From ancient to modern times, the size of ships, the weight of obelisks, the height of cathedrals, the span of bridges, the reach of skyscrapers, the range of spacecraft, the capacity of computers, and the limits of everything have been defined, at least temporarily, by failure. We have built larger and larger as well as smaller and smaller things until we have triggered an alarm, blown a fuse, or reached an impasse in the form of a wake-up call. But size alone is not to blame for failures. Even when we collectively have not been pushing the limits of engineering and technology, we have had rude awakenings. It was not the largest ships at sea that cracked and broke in half during World War II; it was not the longest bridge in the world that was brought down by the wind in 1940; it is not the fastest computer that crashes on our lap in the middle of a routine check of our e-mail. Failures are omnipresent, ubiquitous, mundane, and at the same time both unexpected and ever-expected.

When a highly visible failure occurs, early speculation about its cause often focuses on the design of the broken structure or system. It seems to be a knee-jerk reaction, especially among the mass media, to look for the culprit in the design and its designers. Certainly some failures are attributable to design errors, but they are not the only reason that accidents occur. A design is a manifestation of a technological concept, but a designed thing or system can also be neglected, misused, and mishandled by its owners, managers, operators, and users. Getting to the root cause of a failure can sometimes take years, because it can be so subtle and counterintuitive.

Accidents may occur quickly, but they often follow long periods of normal or near-normal behavior. Boston's late-twentieth-century epic construction project, described as "the most complicated and controversial infrastructure project in American history," had as its primary goal the relocation of the city's so-called central artery from an unsightly and chronically jammed elevated highway structure into tunnels to be concealed beneath new urban park space. The completion of the project was supposed to result in less visible and more efficient movement of traffic. Another goal of the project was to create a separate new vehicular tunnel that would connect the heart of the city with its airport across the harbor. What came to be known collectively as the Big Dig had its political roots in the 1970s; its construction phase stretched over about fifteen years. The $15 billion job was finally declared completed in 2006, but that did not end the saga. Even before the official opening of the various new tunnels to traffic, problems began to arise.[5]

In time, great volumes of water leaked through the concrete walls of the wide, multilane tunnels beneath the city streets. The walls were supposed to be impermeable, and so their failure to be so was clearly unexpected. The cause was traced not to the design of the walls but to the nature of the concrete and, in some cases, to the way it was poured to form them. Since the walls were necessarily thick and deep, large quantities of concrete were required for the job, and one company supplied over 130,000 truckloads of the moldable material, a good portion of which was later found to be inferior, which effectively led to porous walls in places. Also, in some cases, concrete was poured over sand, gravel, clay, and general debris, leading to the inclusion of these materials in the completed walls, which weakened them and made them porous. In September 2004, a section of wall blew out and caused part of the tunnel to flood with ground water that the wall was supposed to keep out. The failure was attributed to, among other things, improperly placed steel reinforcing. In all, about 3,600 leaks large and small had to be repaired, a task that was expected to take as long as a decade.[6]

The problems with the concrete involved not only how but also when it was placed. According to specifications, concrete for various aspects of the project was supposed to be poured in place within an hour and a half of being mixed. However, one supplier admitted that it "recycled old, adulterated concrete" by adding water and other ingredients "to make old loads look fresh." The fraud was further compounded by the falsification of records indicating when the mixing occurred. The magnitude of the offense led to a $50 million settlement being paid by New England's largest concrete and asphalt company. Few designs can tolerate such blatant abuse without revealing unintended weaknesses and downright failures, sometimes years down the line.[7]

The leaking tunnels produced more of an aesthetic and traffic annoyance than a life-threatening condition, but an unrelated and undetected problem in an airport tunnel proved to be fatal. In order to keep noxious gases from accumulating in vehicular tunnels, large quantities of fresh air must constantly be fed in and equally large quantities of bad air exhausted. This process requires what are effectively large plenums, which play the role that air ducts do in a house or building. In a feeder to the Ted Williams Tunnel, which carries traffic between Boston proper and its Logan Airport, the exhaust plenum was formed at the top of the tunnel by hanging a series of heavy (three-ton) concrete panels from the roof. The seemingly dangerous weight of the overhead panels was by design, so that the moving air stream would not cause the panels to vibrate excessively and so create more noise in addition to the already loud traffic. Another reason for the concrete panels was by redesign, since the original plan was to use lighter metal panels with a porcelain finish, but they were more expensive and the project was looking for ways to cut costs. In any case, to keep the overhead panels in place, workers hung them from the threaded ends of steel rods embedded into the concrete roof of the tunnel. The process by which each rod was to be installed consisted of drilling a hole in the concrete ceiling, reaming out the hole with a wire brush to leave a clean but rough surface, injecting an epoxy adhesive, inserting the rod into the epoxy, and holding it in place until the epoxy set. When the rods were firmly installed, the heavy panels were lifted up and bolted in place. On the fatal night, without warning, four of these hung panels came loose and one section fell onto a car carrying a husband and wife to the airport. The car was crushed, the man was severely injured, and the woman was killed.[8]

In the immediate aftermath of the failure, the NTSB began to look into what caused it. Early speculation included such factors as the tunnel temperature, which might have softened the epoxy; and vibrations from a nearby construction project, which might have worked the bolts loose. The design of the panel support system in general naturally came in for close scrutiny. But the system of support was not an untried technology, and so the focus turned to the way the support rods were installed. The use of diamond-tipped drill bits became suspect, since they produce a smooth hole. A hole with a rougher surface, as would be produced by using carbide drill bits, would have resulted in a better bond. Perhaps the concrete dust was not sufficiently cleaned out of newly drilled holes before epoxy was injected into them. This also would have resulted in an inferior bond, and the rod could eventually work itself loose, or so the speculation went.[9]

It finally became clear what the real problem was. Two types of epoxy were manufactured and sold: one was quick-setting and the other was not. Both types came in similar containers, distinguished mainly by the color of the label. The quick-setting epoxy obviously had the advantage of promising faster installation, but it did not have the long-term holding power that the standard epoxy has. Over time, the weight of the heavy ceiling panels very slowly pulled the rods out of the holes, a structural phenomenon known as "creep." The rods holding the fallen panels were pulled out first, but subsequent inspection of the tunnel revealed the phenomenon at work at other locations also, thus confirming the cause. Whether the correct or incorrect epoxy was supplied and used became a subject of debate, but the failure alerted everyone to the potential for confusing different epoxies and their capabilities to resist large pulling forces. There was nothing fundamentally wrong with the basic and much-maligned design of the tunnel ceiling system.[10]

Nevertheless, the structural design of tunnels, buildings, and bridges typically takes place in the relatively clean and calm environment of an engineering office. If there is dust to be dealt with, it tends to be on the surface of a computer screen, attracted and held there by static electricity. Engineers at computers do not expect the strength or integrity of the wall or column or beam or rod whose size they calculate and specify on digital drawings to be compromised by dust or debris on the construction site. But even though everyone knows that conditions out in the real world are not antiseptic and that the work is not free of distractions or outright mistakes, there is the expectation of reasonable care and due diligence in preparing for and placing concrete and the rods anchored into it. Concrete is such a common material that we tend to give it little thought or respect. It begins as what has been described as a "stew" of the binding agent known as cement, to which have been added varying amounts of sand, gravel, water, and other materials. The stew, which must be delivered to the construction site within about ninety minutes of having been mixed, is the stuff we walk on and park our cars on. Whatever is on the bottoms of our shoes gets scraped onto the concrete; whatever oil or other fluid that drips from the underside of our vehicles onto the concrete gets stuck to our shoes. And so it goes. Concrete is such a ubiquitous material that we give it little thought or respect—until it develops cracks or leaks or discoloration.[11]

Yet concrete has been used to build some of the most beautiful structures in the world. The famous ocular dome atop the second-century Roman Pantheon is made of concrete. Robert Maillart's Salginatobel Bridge, completed in 1930 in rural Switzerland, is a masterpiece of the shallow-arch form. The main terminal building at Washington Dulles Airport and the original TWA terminal at New York's JFK Airport, each designed by the Finnish-American architect Eero Saarinen and dating from 1962, are paragons of concrete-shell construction. But for all its ability to be sculpted into works of structural art, concrete often has been viewed as a poor cousin to steel. Very tall buildings have traditionally been steel rather than concrete structures, though that is changing. When Malaysia wanted to erect the tallest building in the world, concrete was favored over steel because the developing country had no indigenous steel industry. In the daring decision to require a concrete structure, engineers designing the Petronas Towers, located in Kuala Lumpur, were challenged to oversee the development of a high-strength concrete and devise a means to pump it up to higher heights than had theretofore been achieved. A decade later, Dubai was a growing municipality of daring architecture of all kinds. Its Burj Khalifa, which opened in 2010, is basically a concrete structure as high as the concrete could be pumped, which was just under two thousand feet, and then steel for another seven hundred feet, to make the complete structure taller than any other built on Earth.[12]

Concrete and cinder blocks are versatile building materials commonly employed in poorer countries. Roads running out from Lima, Peru, to its more rural areas are lined with incomplete concrete structures, their owners often waiting for the material and financial resources to build another story. Many an enclosed ground-level storefront or living quarters are topped with a crown of concrete columns with bits of reinforcing steel reaching out of them toward what someday may be a finished design. Most of these structures are too small and simple to actually have plans, in the sense of formal blueprints, other than in the minds of their inhabitants. If they are in fact designed, it is in a folk tradition.

In Haiti, an even poorer country, reinforcing steel is a luxury. Concrete blocks are often homemade "in people's backyards and dried out in the sun," like mud bricks. Cement, the essential ingredient, is relatively costly, and so concrete mixes are diluted with water and applied thinly to make the material stretch. Furthermore, the quarry sand commonly used in the mix is inferior to river sand. Building blocks made from such materials are very soft and brittle and so easily scratched and broken. In much of pre-earthquake Haitian construction, walls consisted of concrete blocks stacked loosely between concrete columns and with very little mortar in the joints. Finishing consisted of a thin wash of weak concrete. Such construction offered little resistance to being shaken side to side, as the ground has been wont to do on occasion. Moreover, many of the buildings were effectively built on stilts with no walls between them, to allow for car parking on the ground level, and this left the structures "soft" and not able to stand up to the shaking of the earth. Haitians had, in effect, been living in houses of cards built on flimsy card-table legs.[13]

When the 2010 earthquake struck, most such Haitian structures came crumbling down. Six months after the earthquake, as much as twenty-five million cubic yards of debris remained on the ground in Port-au-Prince, and experts estimated that as many as three years would be required to remove it. Haitian survivors had salvaged any building blocks that were not broken and any steel that could be pulled from the rubble of crumbled concrete. With the steel rods straightened, these materials were resold for use in new construction, even though straightened steel would be less strong than it had been in its original state. Experts warned that rebuilding the country with old materials and methods would leave it even more vulnerable to the next earthquake. There were calls against premature rebuilding, and calls for the institution of building codes that might incorporate rules of good design. The use of quarry sand in making concrete blocks was banned. But when so much of the population is living in tents, such calls can seem like a design luxury.[14]

Even if a building is well designed structurally, it can still succumb to failure through no fault of its own. Shanghai is one of China's showcase cities, boasting world-class super-tall skyscrapers. It is also a city of many people, who necessarily all need places in which to live. One cluster of thirteen-story apartment buildings was under construction at a convenient location only five metro stops from downtown Shanghai, and office workers were willing to pay a premium price for that access. Early one summer morning in 2009, one of the buildings began to tilt over and soon collapsed all in a piece onto its side, with only cracks in its concrete walls and façade suggesting the structural trauma it had suffered. The shaking caused by the building's slapping down onto the ground made people in neighboring apartment buildings think an earthquake had occurred. Just a year earlier, an earthquake had caused almost seven thousand school rooms in Sichuan Province to collapse, resulting in the death of more than five thousand children and teachers. According to one report, "unsafe construction is a chronic problem in China and is often blamed on poor planning, shoddy work or theft of materials."[15]

In the Shanghai incident there had been no earthquake, and no flaw was found in the fundamental design or construction of the building itself. The trouble was with some earth-moving work associated with the project. During the main phase of construction, the ground surrounding the building was left flat for convenience. However, after the apartment building was completed, workers began to excavate a deep hole on one side of it to make an underground parking garage. The excavated dirt was carried around the building and piled on the ground on the other side. The asymmetrical condition of a thirty-foot-high pile of heavy dirt on one side and a gaping hole on the other meant uneven sideways pressure on the ground beneath the building. During the time that this condition existed, heavy rains saturated the soil, including that beneath the apartment building. The result was an uneven lateral pressure on the structure's concrete piles, something that they were not designed to resist. When the piles began to break, there was not enough foundation support to keep the building from sliding toward the open pit, and consequently tipping over. It was the design of the excavation procedure and not that of the building itself that was the cause of its failure. Had its designers known that the piles would be subject to sideways pressure, they would have made them larger and thus more resistant, but the design and construction of the building and its garage were evidently not seen as parts of an integrally related system.[16]

Even a flawless design is only as good as the materials used to execute it and the care taken to protect it. As experiences with the Big Dig made abundantly clear, even good concrete handled ineptly can be associated with failures. What then is the hope for inferior concrete? To ensure that bad material does not find its way into construction projects, inspectors on site customarily scrutinize what is provided. When a truckload of wet concrete is delivered to a job site, samples of it are supposed to be tested. Small amounts of the batch can be molded into the shape of a truncated cone and observed for the amount of "slump" they exhibit. This is an early indicator of the material's consistency and suitability. Some other concrete is poured into cylindrical molds, allowed to cure at the construction site, and tested for strength typically at intervals of fourteen, twenty-eight, and fifty-six days. Most concrete achieves a high percentage of its ultimate strength within about a week, and so the tests are a reliable indicator of what will be its permanent behavior. Unfortunately, even something as elementary as testing concrete can be abused in the human system that connects design and construction.[17]

Just as a medical doctor sends patients' blood samples to a laboratory that has the appropriate testing equipment and technical staff, so do construction engineers send concrete samples to a laboratory equipped with machines and lab staff capable of determining the material's strength. The integrity of this routine work should be taken for granted, but some years ago incidents in the New York City area revealed that it was not necessarily wise to do so. A grand jury found that Testwell Laboratories, the city's largest materials testing operation, had over the course of five years "failed to perform tests, falsified reports and double-billed clients for field work." In addition, the company misrepresented the certification of its inspectors. This was more than scandalous behavior, for some of the concrete involved had been incorporated into structures like the Freedom Tower—later to be referred to as

1 World Trade Center—then under construction at Ground Zero, the site of the September 11, 2001, terrorist attacks. In this case, Testwell had certified that the concrete used could support twelve thousand pounds per square inch, the strength specified in the structure's design documents. However, when the building's owner, the Port Authority of New York and New Jersey, later had independent tests performed on the material, it proved to have a strength of only nine thousand pounds per square inch. Since structural engineers base their design calculations on the specifications, their assessment of the strength of the building would have been as much as a third higher than it in fact was. To allow for such a discrepancy engineers employ a factor of safety, thus making a structure as built stronger by that factor than it absolutely needs to be.[18]

Fraudulent test results were also associated with concrete used in Manhattan's Second Avenue subway project and in the new Yankee Stadium, which at the time was being built next to the old one in the Bronx. Testwell was found to have been engaging in "separate schemes involving mix design, field tests, compressive strength, steel inspection and certification" and was charged with "enterprise corruption," which is equivalent to a charge of racketeering. Among the unethical, deceptive, and illegal practices uncovered during the trial was the issuance of hundreds of "phony" reports based on the half-dozen or so mix-design tests that were actually performed. The company's president, V. Reddy Kancharla, was found guilty of falsifying business records, among other charges, and was given a prison sentence of seven to twenty-one years.[19]

When the new Yankee Stadium was being built, construction workers who were also Red Sox fans threw a Boston Red Sox jersey into some concrete being poured, apparently hoping to place a curse on the new baseball field. Fortunately, some other workers, who were more sympathetic to the New York team, eventually revealed the location of the jinx jersey so that it could be jackhammered out of the two-and-a-half feet of hardened concrete that covered it. A possibly more serious problem surfaced at the end of the first season of play in the new ballpark. The pedestrian ramps that fans used to get from one level to another in the stadium developed numerous cracks, some several feet long and an inch wide, large enough for a woman's stiletto heel to become stuck in. While those wishing to downplay the walkway cracking described it as "cosmetic," others noted that the concrete mix was designed and certified by Testwell. An investigation was initiated to determine if the cracks were due to "the installation, the design, the concrete or other factors."[20]

Unscrupulous suppliers and inspectors have long plagued the construction industry. When the Brooklyn Bridge was being built, its board of directors resolved that no contract could go to a firm with ties to any of the trustees or engineers involved in the project. This meant that the Roebling company was not able to supply the steel wire that would go into making the suspension cables, because the bridge had been the design of John Roebling, and after his death his son Washington Roebling had become its chief engineer. At the same time, the younger Roebling unsuccessfully warned the trustees against letting the contract for the steel wire for the main suspension cables go to the company of J. Lloyd Haigh, who was known to be untrustworthy, to say the least. Nevertheless, that company did get the contract, and in fulfilling it, it was later learned, Haigh's firm had indeed perpetrated a great fraud. According to Roebling, "An engineer who has not been educated as a spy or detective is no match for a rascal."[21]

Because Roebling did not believe that Haigh's wire was consistently good, each spool of it had to be tested for strength before being accepted for use in the bridge. Wire that passed the test was given a certificate of approval and sent on to the construction site; inferior wire was rejected and sent back. After some time, however, suspicions arose that some bad wire was finding its way into the bridge cables. The suspicion was checked by making a secret mark on approved wire, and then looking for that mark before using the material. When wire began reaching the bridge without the mark, it was clear that something was amiss. What should have been a steadily growing accumulation of rejected wire was in fact a shrinking amount. It turned out that between inspection and shipment to the bridge, bad wire was substituted for good and sent on with its certificate to be incorporated into the bridge cables. When eighty spools of yet unused wire at the bridge site were tested, only five were found to be acceptable. Before the fraud was uncovered, a total of 221 tons of rejected wire had evidently been used as if it were good. This clearly meant that the strength of the cables had been compromised, and Roebling had to decide what to do. Finding and removing the inferior wire would have been very difficult and time consuming. Fortunately, precisely to allow "for any possible imperfection in the manufacture of the cables," Roebling had designed them with a factor of safety of 6, meaning that they were designed to be six times as strong as they had to be. Roebling decided to leave the bad wire in the cables, recognizing that this would lower the factor of safety toward 5, which would be acceptable as long as no more inferior wire was used. As an extra safety margin, Roebling added to the cables about 150 good wires beyond those originally planned. That the bridge has stood for over a century and a quarter is a vindication of his good judgment. Even with known flaws incorporated into it, a sound design can prevail. It is also the case that a poorly controlled supply chain can compromise quality and integrity and thus lead to failure.[22]

All kinds of inferior materials can jeopardize an otherwise sound design, whether it be of a sports stadium, a suspension bridge, or a private home. At the height of the housing boom in America, before the bubble's burst began to be felt in 2008, there was so much home construction that builders and contractors were having difficulty finding supplies of domestically produced drywall. The shortage led to the increased use of wallboard made in China, with about seven million sheets of it being imported into the United States. The foreign material was incorporated into tens of thousands of homes before problems began to be noticed. Homeowners were complaining of nosebleeds, headaches, breathing difficulties, corroding metal, failing appliances, and a generally foul odor pervading their residences. Sulfur compounds emanating from the Chinese drywall were eventually blamed, and numerous lawsuits resulted. Unfortunately, the vast majority of the imported wallboard could not be traced to its manufacturer, and so no ultimately responsible party could be identified. Many of the affected homeowners—thousands of homes in thirty-eight states were involved—loved the design of their castle, but they could not bear to live in it. The only remedy seemed to be to rip out all the bad drywall and replace it with good, but this could cost in excess of $100,000 per home. According to a spokesperson for the National Association of Home Builders, the dilemma "couldn't have come at a worse time for the industry." Housing prices in general were plummeting, insurance companies were not paying claims on bad drywall, and Chinese manufacturers were not cooperating.[23]

Not only fully formed imported products can ruin what should last a lifetime. The use of polyvinyl chloride piping has become increasingly common in water-supply systems. The plastic material known as PVC has the advantages of being relatively lightweight, easy to cut and install, and corrosion resistant. However, as early as the mid-1990s substandard pipe had found its way into the supply chain, and in some installations PVC pipes that were expected to last for fifty years had been bursting before they were in the ground for a single year. One place where this occurred was in Nevada, where a water main supplying a prison repeatedly ruptured under the pressure of the water it was designed to carry.

Each occurrence required the water to be shut off and the broken section dug up and replaced at no little cost. The affected pipe had been supplied by a single manufacturer, which a whistleblower accused of "falsifying test results about the quality of its products." According to the former employee, the problems could be traced to the introduction of cost-cutting measures by the company. Reportedly it had started using a "lower grade of raw materials" from abroad and had increased the rate of manufacturing output. An industry "honor system" required the firm to submit new samples to the independent Underwriters Laboratories (UL) for recertification and the right to continue to use the UL quality-certifying mark on its products. As might have been expected, the company defended the quality of its products and blamed the bursting pipes in Nevada and elsewhere on the design and construction of the water systems. However, in the face of legal challenges, the company announced that it would guarantee some of its pipe products for fifty years rather than the one year that it had previously offered.[24]

The whistleblower had additionally claimed that the company also cut costs by "hiring people like him straight out of college." Among his first assignments was to field customer complaints, and he reportedly stated that he "was trained to look for ways to attribute leaks and ruptures to the governments and contractors who installed and maintained the pipes." The young engineer was also charged with overseeing tests, and exposure to both the complaint and the testing end of the business led him to put two and two together. He found it increasingly difficult "to tell customers their problems were the result of their own errors on construction sites." This is effectively what the Nevada Public Works Board manager was told when he sought the company's help in determining why the pipes carrying water to the prison kept breaking. According to the works board manager, company specialists who came to Nevada to assess the situation speculated that the pipe was not properly supported and that the contractor who installed it may have "caused a stress on the pipe that it wasn't designed for." This and other aspects of the case were destined to be debated in the courtroom for years to come.[25]

Whether legitimately or not, most if not all failures are initially or ultimately blamed on design in one form or another. Even when some obvious human error or deliberate act of terrorism is identified as the root cause of an accident or disaster, design is inevitably indicted. Whenever there is an airplane crash, the design of the aircraft or its control system is called into question, and sometimes rightly so. If, indeed, there is some generic fault in the type of aircraft that crashed, then all similarly designed planes should be considered likely candidates to repeat the tragedy. Even though the 2001 attacks on the World Trade Center in New York were obvious acts of terrorism, the design of the twin towers was called into question because the structural floor system was said not to be robust enough to withstand a prolonged fire. Furthermore, it was claimed, had the structures been laid out differently, many people who were trapped on the upper stories might have escaped through a stairwell in a location remote from the area of major damage.

But design does not occur in a technological or political vacuum. Engineers are generally responsible for conceiving, evaluating, comparing, and recommending the concept and details of a structure or system; but engineers alone do not make final decisions or choices, for these depend upon more than technical factors. Questions relating to cost, risk, and other economic, social, and political considerations can dominate the decision-making process and push to the background technical details on which a project's ultimate success or failure may truly depend. Although Hurricane Katrina was clearly a natural occurrence, it was the human design and maintenance of the levees and other storm-protection systems around New Orleans that were found wanting. In the wake of the tragedy, even the design of the city's emergency evacuation plans was rightly called into question. New Orleans may not have been devastated by Katrina had the levees and other flood-control devices been designed, built, and maintained to more demanding standards, and the death toll may not have been as great had an adequate evacuation plan been implemented. The compromised technical design of the protective system was certainly not up to the challenge of the hurricane, nor was the design of the city's response to an emergency. Neither design was based purely on technical considerations—no design ever really is—and therein lay the blame.

At the 2010 Winter Olympic Games, many nontechnical factors affected the design of the course that luge and bobsledding events would share. Even before Vancouver, British Columbia, made its bid for the Games, the local organizing committee consulted with the International Luge Federation and the International Federation of Bobsleigh and Tobogganing, which are responsible for track standards. An obvious location for the course was at North Vancouver's Grouse Mountain, and it was an early choice. However, that location was known to sometimes experience "warmish, wet winters that could lead to mushier, slower ice" and hence slower racing speeds. Relocating the course high in the Canadian Rockies, at the ski resort of Whistler, had the added advantage of attracting year-round use that would make the facility "financially viable after the Games." Unfortunately, ideal tracts of land were not available at Whistler, and so the Olympic planning people had to settle for a steep and narrow area, which they knew would mean that any course laid out on it would be "unusually difficult and fast."[26]

The detailed layout and contours of the course, which consisted of a foundation of sculpted concrete beneath the snow and ice, were left to a professional designer based in Germany. Udo Gurgel had more than four decades of experience designing luge courses throughout the world, and he knew that the terrain at Whistler presented exceptional constraints and special design challenges. The extreme narrowness of the location meant that there was no room to incorporate the usual long arcing curves that served to slow down competitors and give them a chance to make adjustments in their trajectory. Because curves at Whistler would have to be short and tight, following them would induce large centripetal accelerations. The associated g-forces would be physically limiting near the bottom of the track, where speeds would be extremely high, and so the lower part of the course at Whistler was almost curveless, meaning that lugers would likely break speed records. Gurgel's prediction of top speeds in excess of ninety miles per hour was confirmed by early test runs completed about two years before the opening of the Olympics. Since he expected top speeds to increase further once athletes got used to the track, he soon revised his estimate to over ninety-five miles per hour and, later, to in excess of one hundred miles per hour. Customarily, in addition to the concrete track, Gurgel would have designed safety features for the course, but Vancouver organizers contracted those out separately.[27]

Serious concerns were raised early on about the safety of the track, and the bobsleigh federation approved it only with the conditions that safety walls be erected and that inexperienced riders be required to start lower down on the course, thereby limiting their top speed. Professional lugers worried about the difficulty of making corrections along the course owing to its paucity of turns and excessively high speeds. Such concerns led the luge federation to recommend that eighty-seven miles per hour should be the top speed on future track designs. Unfortunately, that theoretical speed limit did not apply to Whistler or other existing courses. Moreover, to have imposed such a limit on Vancouver Olympians would no doubt have raised objections, for speed is at the same time the attraction and the danger of the sport, and at Whistler "the course's dangers became part of its marketing."[28]

The various Olympic committees involved in the selection of Whistler as the location for the luge course were concerned about such things as the weather there, the logistics of transporting Olympians to and from the venue, and the financial soundness of the choice during and after the games. The international federations would have been concerned about any course anywhere meeting their standards. Designer Gurgel was presented with what he knew to be a very challenging design problem, but professionals are used to being challenged. His computer simulation of a luger hurtling down the tightly constrained track had predicted that it would be fast, but when test runs proved the course to be even faster than the simulations, decisions had to be made. After some early high-speed accidents on the course, especially surrounding its Turn 12, the ice profile there was altered before a 2009 World Cup event. However, the track remained "as fast and difficult as any in the world." The International Luge Federation demanded further changes, which led to the erection of more safety walls.[29]

Unfortunately, the modifications did not reach down to Turn 16. It was there that Nodar Kumaritashvili, a young luge racer from the Republic of Georgia who was taking a practice run just hours before the Games' opening ceremony, lost control on the extra-fast course and was thrown off his sled and against an unpadded steel pole. It was natural to ask whether the design of the track was to blame for his death. And, if so, who was to blame for the design?[30]

According to the president of the bobsleigh federation, on reporting the accident "the Royal Canadian Mounted Police did not attribute it to design flaws and did not attribute it to speed." Indeed, according to the former bobsledder, "the reason they call it an accident is that nobody can define the cause." This may be true in the immediate wake of a fatal crash, but accidents of such consequence usually lead to further investigation, not only to try to get to the cause but also to be able to apply the lessons learned from such an event to prevent similar accidents in the future. Certainly the Vancouver Olympic Committee and other vested interests did not wish to see a tragic accident cast a pall over the Games for which they had planned for so long, and so it was natural that initial assessments would have been dismissive of any reported design flaw in the course. One thing that is likely to survive the closest scrutiny, however, is that there was no deliberate conspiracy among Olympic organizers, regulators, and designers to kill a young Georgian just hours before opening ceremonies.[31]

Whether it be an airplane crash in freezing rain, a devastating earthquake in one of the world's poorest countries, or an athlete thrown into a pole, an extraordinary event calls for an extraordinary investigation into the cause. There may be as many motives behind the investigation as there are theories about the cause, but under ideal circumstances, if the investigation proceeds without prejudice and runs its course, a final report should identify a most probable cause and a most responsible party. This is not always the outcome, however, because some accidents truly are accidents, with no one decision or no one person to blame. But no matter how devastating a failure, its consequences can be even greater if its lessons are not learned and heeded.

One lesson that can be expected from any investigation is that people were responsible in some way or other, directly or indirectly—even if they are deemed not to have been blameworthy. The difference between being responsible and being blameworthy has to do with involvement and intent. As much as we might read of intelligent machines and self-replicating robots, all technological objects are ultimately the products of human intervention in the natural order of the world. We humans first conceive of and design even the most autonomous systems, and we inadvertently invest our human limitations in them. Except for the likes of the mad scientist determined to conquer the world or the terrorist committed to destroying life, including his own, those people whom we call inventors, designers, and engineers set out to achieve what they perceive to be a good end and to do it with as much care and dedication as they are capable of mustering. The creators, maintainers, and operators of technology are by and large capable and committed individuals and teams who above all else want their creations and plans and charges to succeed. That they some-times fail is but testimony to the humanness of the people in-volved.

Technology is rooted in design, that creative act that exploits the raw materials of the Earth and, working within the tradition of the crafts, draws upon the cumulative knowledge of the arts and sciences and the store of prior technological achievement to bring forth creations of which our forefathers must hardly have dreamed. The wonders of modern technology—and the latest technology is by definition always modern—give our world an ever-changing freshness that is at the same time short-lived, ever to be superseded by the new gadgets and devices that have come so much to characterize the developed societies that the develop-ing ones seek to emulate. On the timescale of such rapid change—and, in the context of the history of technology, even the passage of decades and centuries can seem to be lost in the blink of a ro-bot's electromechanical eye—there can naturally be a confusion of purpose and a loss of concentration on the technological goal, which is success. When this happens, failure cannot be far behind.

Failure can be an inevitable consequence of lapses in attention, judgment, and purpose. Individuals of integrity may not be able to imagine themselves falling into the trap of complacency and deception, but it happens all the time. Especially when we become involved in groups and teams, the moral compass of the individual can be deflected by the mettle of the masses. Insidious influences can lead engineers to ends that are antithetical to their nature and to that of their profession. These deviances are seldom intentional; they are the unintended consequences of the pursuit of a design goal that may be reachable only with such difficulty that its pursuit consumes the team's entire creative energy and so blinds it to tangential concerns, which may prove to be the ones that lead to failure. The surest way to avoid such pitfalls is to be familiar with those that have ensnared our predecessors. This book, for its part, describes in a human context significant and instructive failures from the distant and recent past and seeks an understanding of how and why they have occurred so that mutatis mutandis they might not be repeated.

TWO

Things Happen

It should not surprise us that failures do occur. After all, the structures, machines, and systems of the modern world can be terribly complicated in their design and operation. And the people who conceive, design, build, and interact with these complex things are unquestionably fallible. They sometimes employ faulty logic, inadvertently transpose digits in a numerical calculation, mistakenly overtighten a bolt or undertighten a screw, casually misread a dial, or hurriedly push when they should pull. They also can fail to concentrate, to anticipate, and to communicate at critical moments. At other times, accidents can occur because people cease to be honest, to be ethical, and to be professional. For whatever reason, accidents happen, and accidents invariably lead to or from the failure of something or someone. What should surprise us, really, is not that failures occur but that they do not do so more often. When they do happen on our watch, we tend to defend ourselves against accusations; we try to shift the blame. Our faults are all too often imputed to the things we design, make, sell, and operate, not to the people who design, make, sell, and operate them.

Technology has always been risky business, but quantifying that risk is a relatively new phenomenon in the worlds of engineering and management, which should be more integrated than they often are. The space shuttle program clearly needed large numbers of engineers and managers to accomplish its mission, and for planning purposes it also needed a sense of how successful it could be. Each shuttle consisted of millions of parts, which only suggested the degree of complexity of the entire system of hardware, software, and operations. In the early 1980s, managers at the National Aeronautics and Space Administration (NASA) estimated that the flights would be 99.999 percent reliable, which represents a failure rate of only 1 in 100,000. According to the physicist Richard Feynman, who was a member of the commission that investigated the January 1986 *Challenger* accident, in which the shuttle broke apart shortly into its flight, killing all seven astronauts on board, this "would imply that one could put a Shuttle up each day for 300 years expecting to lose only one." He wondered, "What is the cause of management's fantastic faith in the machinery?" Engineers, who were more familiar with the shuttle itself and with machines in general, predicted only a 99 percent success rate, or a failure every 100 launches. A range safety officer, who personally observed test firings during the developmental phase of the rocket motors, expected a failure rate of 1 in 25. The *Challenger* accident proved that estimate to be the actual failure rate, giving a success rate of 96 percent after exactly 25 launchings.[1]

The failure of *Challenger* understandably led to a rethinking of the shuttle's design details and operation, and changes were made on the basis of lessons learned. After a twenty-month hiatus, missions resumed and the shuttle fleet flew successfully until the 113th mission, which ended in *Columbia*'s disintegration upon reentry into the Earth's atmosphere in 2003. The historical record then proved the success rate, which had been at 99.11 percent just before *Columbia*, to be 98.23 percent. This figure increased to 98.48 percent as of May 2010, when *Atlantis* returned from its final scheduled flight. This left the space shuttle program with only two remaining planned flights, and with their completion the success rate the program achieved was 98.51 percent, short of even the engineers' prediction. According to a minority report from a group that had monitored progress in shuttle safety after the *Columbia* accident, managers at NASA lacked "the crucial ability to accurately evaluate how much or how little risk is associated with their decisions." No matter what the technology is, our best estimates of its success tend to be overly optimistic.[2]

Indeed, "We were lucky" was the way NASA summarized the results of a shuttle program retrospective risk assessment released in early 2011. The chance of a catastrophic failure occurring in the first nine shuttle missions was in fact as high as 1 in 9, representing a success rate of less than 89 percent. In the next sixteen missions, which included the *Challenger* mission of 1986, the odds of a failure were 1 in 10. The odds changed throughout the program because modifications to the system were constantly being made. For example, when the Environmental Protection Agency (EPA) banned the use of Freon, NASA had to stop using it to blow insulating foam on the external fuel tank. The compound used to replace Freon did not allow foam to adhere as well to the tank, resulting in more foam being shed during liftoff and flight. This increased the risk of an accident, such as the one that would eventually destroy the shuttle *Columbia*. For the nine shuttle missions that were flown in the wake of the Freon ban, the odds of a disaster increased from 1 in 38 to 1 in 21.[3]

Of course, engineering and technology are not spectator sports, judged by the final score. Preparing and launching a space shuttle involved many teams, which were expected to work in concert rather than in competition with one another. The teams had a single objective: the successful completion of each mission, from which the aggregate record would follow. The opponent, so to speak, was not another team or set of teams—although it was the Soviet Union in the case of the race to the Moon—but nature and nature's laws, of which the eighteenth-century poet Alexander Pope wrote in an epitaph intended for Isaac Newton:

Nature and Nature's laws lay hid in night:
God said, Let Newton be! and all was light.[4]

As much as he was lionized, Newton himself realized that he was but part of a team, comprising perhaps some contemporaries but most importantly predecessor colleagues in mind and spirit who had wondered about the same mysteries of the universe as he. As Newton wrote in a letter to his scientific contemporary Robert Hooke, "If I have seen further it is by standing on the shoulders of Giants." We all stand on the shoulders of giants who preceded us in our continuing quests for whatever is forever to be achieved beyond the horizon. In engineering the holy grail is the perfect design, something that always functions exactly as intended and that never needs any improvement. Of course, if we could achieve it, the perfect design would never fail.[5]

For Newton, all may have been light, but it was also heavy. The struggle of the space shuttle against the force of gravity was evident in the agonizingly slow early seconds of liftoff during each launch from Cape Canaveral. Of course, once the struggle had been won, gravity became an ally, keeping the shuttle in low Earth orbit even as it wanted to follow its velocity off on a tangent. When the space age had dawned in the second half of the twentieth century, the basic physical laws necessary to design and fly spacecraft were believed to have been more or less fully illuminated. Otherwise, manned flights into orbit and beyond would have been a much riskier endeavor, if not just a fanciful dream. The trick was to exploit the laws properly. But just knowing the laws of nature is not sufficient to field a team to compete successfully against them. It takes the creative genius of engineering to design a spacecraft like the shuttle that will not only be launched successfully but also orbit Earth, reenter the atmosphere, and glide to a safe landing. Success demanded the integration of a great amount of specialized knowledge and achievement by teams of engineers engaged in the intricacies of rockets, combustion, structures, aerodynamics, life support, heat transfer, computer control, and a host of other specialties. Each member of each team had to contribute to the whole effort. There had to be give and take among the teams to be sure that no aspect of their singular goal worked at cross purposes to another.

In any project, large or small, each engineer's work is expected to be consistent and transparent so that another engineer can check it—by following its assumptions, logic, and computations—for inadvertent errors. This constitutes the epitome of team play, and it is the give and take of concepts and calculations among engineers working on a project that make it successful. Of course, slips of logic do occasionally occur, mistakes are made and missed, resulting in a flawed design, which may or may not lead to an immediate failure. If the project involves a building, for example, an underdesigned beam or column might reveal itself during construction. It might bend noticeably and so not look quite right to a field engineer's trained eye, which might send the designer back to the drawing board, where the error might be caught. Unfortunately, not all errors are caught, either in the design office or on the construction site, and those that are not can indeed lead to failures.

Parking decks are familiar structures that do fail now and then, and the failures can often be traced to something out of the ordinary in their design or construction. Such collapses might never have occurred if the structures and everything surrounding them had been exact copies of those that had stood the test of time, but even repeated success is no guarantee against future failure. In fact, prolonged success, whether it be in a space shuttle program or in the design and construction of parking garages, tends to lead to either complacency or change, both of which can ultimately lead to failure. As one engineer has put it, "every success sows the seeds of failure. Success makes you overconfident." When we are overconfident and complacent, satisfied that we have been doing everything correctly because we have had no failures, we also tend to become inattentive and careless. We begin to take chances, and good luck like the kind that was had in launching shuttles with faulty O-rings runs out. Or, if we are experiencing a string of successful projects involving parking garages, say, we begin to think that we can make them a bit more competitive by using lighter beams or by introducing a more efficient construction technique. Then, the structural flaws that had lain hidden from sight can become revealed in the collapse that lets in light.[6]

In the spring of 2010, the oil well blowout that led to the explosion on the *Deepwater Horizon* drill rig, its sinking, and the subsequent prolonged oil leak in the Gulf of Mexico took everyone by surprise in part because few remembered that anything quite like it had ever happened in the area. But in fact, three decades earlier, in 1979, the *Ixtoc I* exploratory well that was being drilled by a semisubmersible rig operating in little more than 150 feet of water experienced a loss of confining pressure, and the subsequent blowout continued to leak oil over the course of almost a year. Ultimately more than three million barrels of crude oil gushed into the Mexican waters of the Gulf and beyond. In the immediate wake of that accident, the oil industry operated with a heightened awareness of the possibility of well failure, and so took extra precautions and more care with operations. Over time, however, and with a growing record of successful drilling for and extraction of oil from Gulf waters, oil rig and well operations grew lax, and this produced the kind of climate that set the stage for the *Deepwater Horizon* explosion and subsequent environmental catastrophe. It was no accident that these two unfortunate events occurred about thirty years apart, for that is about the span of an engineering generation and of the technological memory for any industry. During such a career span, we can expect periods of success punctuated by incidents of failure, and depending when in the cycle a young engineer enters the industry, he or she can be more sensitized by one or the other. This sensitization tends to dominate design and operational behavior for a period, but in time a paradigm of success tends to suppress one of failure, and an atmosphere of overconfidence, complacency, laxity, and hubris prevails until a new failure provides a new wakeup call.[7]

The cyclical nature of success and failure has been well established in the field of modern bridge design and engineering, in which experience spans about two centuries. Unfortunately, the lessons learned from failures are too often forgotten in the course of the renewed period of success that takes place in the context of technological advance. This masks the underlying fact that the design process now is fundamentally the same as the design process thirty, three hundred, even three thousand years ago. The creative and inherently human process of design, upon which all technological development depends, is in effect timeless. What this means, in part, is that the same cognitive mistakes that were made three thousand, three hundred, or thirty years ago can be made again today, and can be expected to be made indefinitely into the future. Failures are part of the technological condition.[8]

We might hope to achieve a world with fewer failures if a moratorium on invention and innovation were declared. This would necessarily freeze technology at its present stage. Everything made subsequently would have to be designed and produced exactly like what had already been demonstrated to have been successful. But a static technological architecture would result. We could expect no improvements in automobiles from year to year. Computers and other personal electronic devices would not evolve in performance or price. Under such conditions the world would certainly be a safer place, but the joy of living would be somewhat diminished. Deprived of technological change, people could be expected to suffer from the equivalent of cabin fever, when the winter weather remains unchanged for a prolonged period of time. Maybe then they would engage in riskier behavior on the highway, just for a change, or engage in computer hacking, just for a change.

Freezing technological innovation would also likely have profound effects on the economy. Without innovation, there would be no market for new products. People would likely continue to drive their old cars, because new ones would not be fundamentally different, having no new features, safety or otherwise. New housing sales would likely suffer for the same reason. Without the attraction of new consumer goods, some people at least would not be motivated to work or save for a new television or entertainment center, for there would be nothing new. Society would stagnate.

Change is an essential part of the human condition, a fact implicit in the evolution of civilization and culture. This has been explicitly acknowledged since ancient times, when the Sphinx posed the riddle, "What creature walks in the morning on four feet, at noon upon two, and at evening upon three?" It was Oedipus who solved the riddle, whose answer is, of course, men and women. We crawl before we walk, and we eventually need a cane when our two legs are no longer strong enough to support us in our wobbly old age. Aging and changing—and the technological innovations of walking canes and more—are just part of being human.[9]

The history of technology is a history of change, though not necessarily rapid change. In *The Ten Books on Architecture*, the first-century BCE volume that is considered the oldest extant written work on architecture—a field that then included what we today call engineering—Vitruvius laid down rules about how classical buildings should be planned, oriented, and proportioned. These prescriptions did not totally suppress creativity, but they severely limited its range. That is not to say that classical architecture did not produce masterpieces that remain so today. Indeed, the rigidity of classical styles—and the proscriptive nature of contemporary construction—may in part be responsible for the survival of the masterpieces. In Vitruvius's time, architecture and the engineering that realized its designs were essentially two sides of the same technological coin. Concerns about aesthetics and buildability were often inseparable. Indeed, one reason that Greek temples were characterized, relatively speaking, by closely spaced columns was that the classically proportioned stone architraves that span the distance horizontally between the adjacent vertical supports could only be made so long before they would crack and break under the burden of their own considerable weight, either during the process of being lifted into place or after the additional weight of friezes and pediments was placed upon them.[10]

In spite of the title of his book, Vitruvius related stories not only of classical architecture in its modern sense but also of evolutionary change in what might be described as construction engineering. In particular, he showed how emulating successful procedures could lead to failure. He described methods for transporting heavy pieces of stone from quarry to building site and explained how the devices employed were successively modified to accommodate new conditions and meet new constraints. Thus, whereas round pieces of marble could be rolled along the ground after being fitted with timber frames to which oxen could be hitched, prismatic architraves could not be handled in this way. To move these heavy rectangular pieces of stone, workers had to treat them like axles between timber wheels built up around their ends. According to Vitruvius, the scheme was once modified further to transport a more cubic piece of stone that had been prepared to serve as a new base for an old statue. The statue was located in a central place in the city that had grown up around it, and now narrow and curving streets separated the quarry from the destination. To make the journey, the stone was enclosed in a spool-like timber cage, around which a rope was wrapped and attached to oxen. In this way, the moving device was no wider than the stone itself and so could fit between the closest-set buildings along the way. Unfortunately, as Vitruvius related, the scheme ultimately did not work because the path that the spool took could not be controlled as the oxen pulled farther and farther away from their burden. The successive modifications of a scheme that had worked so effectively for rolling marble columns led to failure when adapted to pulling a block of stone.[11]

Into the Renaissance, architecture continued to be dominated by geometric considerations. But in his seminal *Dialogues Concerning Two New Sciences,* Galileo recounted contemporary examples of spontaneous failures associated with moving and erecting large pieces of stone, such as obelisks, and launching large timber ships, especially when they were geometrically greater in size than those previously attempted. Taking a cue from the geometrically dissimilar proportions of the corresponding bones of different animals, he realized that the occurrence of engineering failures indicated that something distinct from geometry had to be considered in scaling up successful structures. He reasoned that this had to be the strength of the material involved and went on to explain how that could be taken into account in the design of beams and, by extension, of structures generally. Galileo's approach, which forms the basis for the analytical methods of modern structural engineering, is a classic example of failures raising questions, the answers to which lead to new successes.[12]

As Galileo's observations made clear, the geometric scaling up of successful structures as a method for designing larger and larger ones, whether they were wooden ships or stone cathedrals, worked only up to a point. That point was reached when the weight of the structure overtook the strength of the material to support it. Before Galileo's insight, the only logical conclusion to draw from the success of a structure was that it was successful. There was no way of determining how close to failure it might have been. By integrating strength and size, Galileo provided a means of inferring success or failure in structures by means of analysis, the foundation of modern engineering. Ironically, Galileo made an unwarranted assumption on the way to reaching his conclusion about a particular beam configuration, which introduced an error that was not discovered until failures began to occur in water pipes designed according to his result. The failures provided incontrovertible evidence that something was amiss, and once that was obvious, the failure could be studied, the error could be uncovered, and a correction could be made. This is the process that begins with what is now known as failure analysis.[13]

Today we understand how important it is to think beyond just the geometry of a structure. We know that we must consider whether it is made of timber or steel or concrete, or any other material. The material matters as much as does the geometry. But not all technological systems are products of structural engineering alone. Indeed, as important as the structure of an airplane may be, it is only one aspect of the system that must be considered to make it a successful overall design. The engines, the instruments, the controls, the crew must be as reliable as the structure if the plane is to take off, fly, and land successfully. And no matter how many times a particular airplane has been flown without incident, that alone is no guarantee that its next flight will also be successful. Hence the importance of preflight checks and scheduled inspections carried out according to protocol, as well as the investigation of anomalies. Anything out of the ordinary could be a signal that the system is not what it was on the last successful flight.

A single failure, by contrast, whether of an airplane or of anything else, is a source of knowledge we might not have gained in any other way. As Vitruvius recounted, the failure of a scheme to move a block of stone revealed in retrospect faulty reasoning that was not obvious in prospect. Engineering design is all about projecting past experience through the lens of present methods into the future. Documented failures are among the most valuable experiences, because they reveal weaknesses in reasoning, knowledge, and performance that all the successful designs may not even hint at. The successful engineer is the one who knows not only what has worked in the past but also what has failed and why.

Even failures that occurred years, decades, centuries, and millennia ago can inform our designs today and prevent them from turning into failures. For example, in his *Dialogues* Galileo tells a tale of unintended consequences involving a long, slender piece of marble that was being held in storage to be used as a monolithic column. The column was resting horizontally across two supports, one near each end, a configuration that engineers today call a simply supported beam. According to Galileo, the situation reminded an observant worker of an obelisk being moved and a ship being launched, both of which were precarious operations that had led to failures. In order to prevent the column from breaking in two like an obelisk or a ship, the worker suggested employing a third support under the center of the piece of marble. Those who were consulted agreed that this was a good idea, and a new support was carefully placed directly under the center of the piece of marble. This led everyone involved to believe that they had eliminated the risk of the column's breaking, and they did not give their improvement a second thought. But some time after this precautionary action was taken, the marble column was discovered broken in two—not by taking the shape of a V between the two supports, as had been feared, but by forming a Λ shape across the new support. Evidently the original end supports had settled into the soft ground, leaving the marble column balanced on the carefully placed middle support. Since the marble's strength was not sufficient to maintain this position, the column broke apart in the middle and its ends dropped to the ground.[14]

Galileo's story of the marble column demonstrates very graphically the general and universal design principle that making any change in a system can introduce a new way in which it can fail. Had the column been left in its original configuration, with the end supports naturally located just short of the physical ends of the column, it would no doubt have survived the gradual settling of the supports. However, introducing the third support, and placing it so carefully right under the exact center of the column, allowed the maximum overhang on either side of the support and thereby the maximum stress to develop in the marble as one or both of the original supports receded from beneath it. When the column broke, its two halves assumed a configuration that would have been impossible had the third support never been added. It was the design change of adding the third support that introduced the possibility of the new failure mode.

A familiarity with the seventeenth-century design change and failure reported by Galileo may have obviated the 1981 collapse of the Kansas City Hyatt Regency Hotel walkways that claimed 114 lives. In that case, the design change of replacing a single continuous steel rod with two offset rods to support a pair of elevated walkways was the cause of the fatal accident. Had the walkway engineers been familiar with the story of Galileo's marble column, and the ill-conceived idea of changing the number of supports, they might have been prompted to consider that making a change from one rod to two could introduce the possibility of a new mode of failure, the realization of which in turn might have caused them to recalculate the suitability of the new connection. Had they done so, they likely would have discovered it to be far from a real improvement, and they would also likely have made some changes in the design of the support system before it was built. Had that hap-

pened, the walkways might still be standing and many of the victims still living today.[15]

Even Vitruvius's two-thousand-year-old recounting of the failure of the scheme to move a block of stone could alert present-day
engineers to the potentially dangerous consequences of making
any changes in a system that has worked. No matter how incrementally evolutionary a contemplated change might seem, it will
have the potential to introduce a new way in which the changed
system can fail. The overarching lesson is that making any change
in a design can alter the entire context in which the detail is embedded and thereby introduce a failure mode that would have
been impossible in the original design. A single change can change
everything.

The engineer in possession of a catalog of failures, however old,
is an engineer equipped with an anthology of horror stories and
an arsenal of arguments that may be called to bear when a colleague suggests even a supposedly minor design change. The seemingly innocuous change, perhaps presented at a design conference,
might appear to be an obvious improvement to the engineer making the suggestion, but to the engineer schooled in failure case
studies it should be a red flag. Describing analogous cases from
millennia, centuries, and decades ago should be sufficient to con-
vince the engineer proposing the change at least to do a failure-
modes analysis or back-of-the-envelope calculation. Even if the
proposing engineer is unconvinced of the potentially negative im-
port of the change, it is possible that a more experienced colleague
at the table might be. And although he may be unlikely to admit
it at first, the proposing engineer himself might have had just
enough doubt raised in his mind to give his design change a second thought, and maybe even do some further analysis on his own
time. He may even come to the next design conference with a
modified change.

Confidence and doubt, like success and failure, are part of the human personality. We all wish to avoid failure and to succeed at what we do, whether it be as a professional engineer or as any other member of the species *Homo faber,* which is characterized by its fabrication of things. Whether knapping flints or splitting atoms, human beings strive to transform things found in the given world into things that help us perform tasks and acquire knowledge that are not part of the natural order of things. *Homo faber* wants to change that order and to do so without failing—and with confidence that it will work.

When someone succeeds in acquiring and perfecting a useful skill, we honor that person as a master. The master's method becomes a model to emulate. This is the way technology was passed down through generations. This is the way effective arrowheads continued to be made in countless numbers, many of which still lie about prairies and plains, waiting to be discovered by weekend archaeologists and artifact hunters. These adventurers know what to look for because although each arrowhead has its own unique facets, it is immediately recognizable as an example of the deliberately designed thing.

We do not knap stones into arrowheads any more, but we do still design and make things of all kinds. Some of those things are computer software packages and some are tall buildings. Whatever they are, they are subject to failing in one way or another. Success and failure can also be relative, in that what may be generally hailed as a successful design might be viewed as a failure by an inventive critic. The 1970s, when the Citicorp (now Citigroup) Center in New York City was in the planning stage, was a time of growing ecological awareness, and solar energy was seen as an environmentally responsible means of producing electric power. Installing solar panels on the inclined south-facing portion of the building's distinctive top may have made not only promising energy sense but also good public relations sense. And the base of the building was to be as distinctive as its top, but for a different reason. When Citicorp was acquiring land for what was to be its flagship structure, St. Peter's Church at the corner of East 54th Street and Lexington Avenue refused to sell its land but was willing to make a deal for air rights, which made it possible for Citicorp to commission a square building whose northwest corner cantilevered out over the new church that Citicorp erected for St. Peter's as part of the deal. In fact, all four corners of the Citicorp tower were to be cantilevered, and this called for some unusual structural framing. With no conventional corner supports, the 915-foot-tall structure rests on pillars that are located at the center of each of its four sides. It is as if the legs of a card table were moved to where the card players usually sit, leaving the corners suspended over thin air—or card players' knees. Such a table would naturally be more easily tipped over than would a conventional one, and it would do so more easily over a pair of adjacent legs than over a pair of opposite ones.[16]

This characteristic almost proved to be the literal downfall of the new Citicorp Building. Any tall building, if it is not stiff enough, can sway excessively in the wind. To counter this problem, a device known as a tuned-mass damper was installed near the top of the Citicorp structure to offset excessive motion. To give the unusually framed building added stiffness, its joints were to be welded. However, somewhere between initial design and construction, bolts were substituted for welds, making the structure more flexible and thereby more susceptible to collapse in a hurricane. This potential disaster was averted by the quick thinking and integrity of the building's structural engineer, William LeMessurier, who as soon as he learned of the design change made some quick calculations and convinced the corporate owner of the recently completed structure to weld reinforcing plates at the bolted joints before the hurricane season began. This was done, and the building has stood safely since 1977.[17]

Not all design errors or changes are so quickly detected or so expeditiously fixed. Software is notorious for containing bugs, which sometimes are removed by updates and sometimes are not. Occasionally an update even introduces new bugs, just as a design change can introduce a new failure mode. Checking a software fix by just testing it in isolation from the entire program in which it will be embedded can be as meaningless as testing a new beam support in isolation from the beam itself. In any case, computer bugs that survive the extermination processes known as software development and beta testing tend to be innocuous ones that hardly matter. That is why they generally go undetected, except in unusual circumstances. In the mid-1990s, a mathematician discovered that the then-new Pentium microchip sometimes produced incorrect results when multiplying two very large numbers together, something the ordinary computer user would not likely have reason to do. At first the mathematician had looked for the error in his own calculations, but he and other scientists and engineers could not track it down. Intel, the chip's manufacturer, finally admitted that a "subtle flaw" in software was embodied in the chip. This all suggests a kind of syllogism: "Technology is made by people. People make mistakes. So, technology is made of mistakes." In other words, all technological systems are potentially full of bugs and gremlins. When the latent errors, faults, and mistakes are innocuous and do not interfere with the system's overall function, they can go undetected for its lifetime. Something that works, no matter how fundamentally flawed it might be, we call a success. When the flaws in a system metastasize and cause it to malfunction and possibly even cease to function entirely, we call it a failure. The design may be blamed, but, to paraphrase the comic-strip character Pogo, the design is us.[18]

The Apple iPhone was notorious among some users for having poor cell-phone reception and dropping calls, and the issue came to a head when the iPhone 4 was released on June 23, 2010. Reports soon began to circulate that the number of bars indicating the strength of the signal would drop off noticeably when the device was held in a certain way. One of the new model's advertised design features was an improved antenna—embodied in the shiny metal band known as a bezel that wrapped around the edges of the device—which some users blamed for the annoying behavior, especially when their fingers or palm were in contact with the lower-left portion of the phone. Apple's early recommendation to users who saw signal strength drop when they held the phone in a certain way was simply to hold it in a different way. Then, in an open letter to iPhone 4 users dated July 2, Apple admitted to being surprised at reports of reception problems. (At least one sociologist uses the word *surprise* "to avoid the negative connotation of the word *failure*, which implies that mistakes have been made.") After noting that the iPhone 4 was "the most successful product launch in Apple's history" and describing it as "the best smartphone ever," the letter stated that the company had begun investigating the reports immediately upon reading them. What it discovered it described as "simple and surprising." Apple admitted that it was "stunned to find that the formula we use to calculate how many bars of signal strength to display is totally wrong." This meant, for example, that when the phone showed four bars, indicating a strong signal, it maybe should have been displaying only two. Thus, a user would think she was getting strong reception when she was not. Apple admitted that the flawed software had been used in iPhone models from the start, but one observer accused Apple of "trying to exculpate themselves by saying this seeming flaw is not a flaw because it has been there for such a long time." However, in its letter Apple announced that it would fix the bug by adopting a formula recommended by AT&T for calculating the number of bars to be displayed.[19]

Why Apple did not discover the flaw in the algorithm used to calculate the number of reception bars displayed on the iPhone before the iPhone 4 was released may never be known for sure, but it is surprising that the company did not do so. Perhaps its earlier failure analysis of poor reception focused exclusively on improving the antenna to the point of missing the more fundamental cause, that of the false indicator of signal strength. *Consumer Reports* did not buy Apple's explanation and maintained that the problem with the iPhone 4 was indeed a hardware issue. Apple did find a quick and dirty fix in covering the troublesome bridgeable gap in the metal antenna with a strip of duct tape, which is nonconducting but clearly defaced the phone's much vaunted appearance. Later, in an update, the consumer champion endorsed a "bumper case"—a partial cover of rubber or plastic—that keeps the antenna from coming into direct contact with the user's palm or fingers. However, the *Consumer Reports* product reviewer did not forgive the design, nor could he recommend the new phone until Apple came up with a free and permanent fix for the problematic antenna. The company did do so, in the form of free bumpers that, it claimed, "add a touch of style to any iPhone 4."[20]

The Apple iPhone 4 issue was further complicated by a news leak that an Apple "senior antenna expert" had warned Steve Jobs, then the corporation's charismatic chief executive officer, early on in the design process that the antenna design was flawed and could lower reception and result in calls being dropped. The engineer, identified as Ruben Caballero, reportedly told Apple's management that the wrap-around antenna design "presented a serious engineering challenge." In order to be able to work for the various radio-frequency ranges employed by different wireless phone networks, the metal band would have to consist of distinct sections that were insulated from one another. The thin nonconducting gap near the lower-left corner of the iPhone 4 that marked the beginning and end of the metal strip forming the bezel could easily be bridged by a human finger, which is a conductor, possibly resulting in a dropped call. According to Jobs, Apple tested the design and "knew that if you gripped it a certain way, bars would go down a little bit," but "didn't think it would be a big problem because every smartphone has this problem." Thus, in fact, the engineer's concerns about the antenna design proved to be well founded. There were also other warnings about the antenna problem, but Apple's management evidently gave a higher priority to favoring a strikingly sleek design that resulted in a "lighter, thinner handset" and to meeting an announced debut date than to releasing a product without a flawed antenna design. In this regard, the Apple iPhone 4 debacle was reminiscent of a culture of acceptance of flaws and of the engineer-manager disagreements embodied in such avoidable catastrophes as the loss of the space shuttle *Challenger* and the explosion and subsequent Gulf of Mexico oil leak associated with the semisubmersible rig *Deepwater Horizon*, which will be discussed in more detail later in this book.[21]

In all cases of surprise or failure, the greater technological tragedy is not having failures but not learning the correct lessons from them. Every failure is a revelation of ignorance, an accidental experiment, a found set of data that contains clues that point back to causes and further back to mistakes that might have been made in design, manufacture, and use. Not to follow the trail to its source is to abandon an opportunity to understand better the nature of the technology and our interaction with it. Because successful design is the anticipation and obviation of failure, every new failure —no matter how seemingly benign—presents a further means toward a fuller understanding of how to achieve a fuller success. There will always be plenty of new mistakes to make, without repeating the old. These mistakes include, of course, not only those involving hardware and software but also those involving engineers and managers and the dropped or ignored communications that result from their insulation from each other. [22]

This is not to say that we should encourage, promote, or cheer on failures of any kind. No one, especially an engineer, wants a system or device to fail to perform the function of its design correctly and completely. But because we are all human and thereby fallible, we should not be so presumptuous to think that anything we design is without a fault that we introduced into it. As Newton observed in the preface to his *Principia Mathematica,* "errors are not in the art but in the artificers." When errors and faults become manifest in an overt failure, we must own up to our role in sowing the seeds of failure in the design and take as much care as possible in deconstructing the accident and broadcasting its lessons. It is only by doing so that we can hope to gain some absolution and help ourselves and our professional colleagues not make the same mistake again. To drive home the lesson, we might paraphrase the dictum about being fooled, imagining the designer to be addressing his design: "Fail me once, shame on you; fail me twice, shame on me."[23]

The surest way to avoid being fooled is to be wise in design, and the route to wisdom and success is through understanding failure. Naturally, the more we know about failure specifically and generally, the better we can hope to understand it. Errors and mistakes leading to failure have been made as long as humans have been conceiving and designing improvements to the world as we find it. However, we continue to make errors, and there is no reason to think that we will not continue to do so into the indefinite future. We are human, but we are also technological animals, and as such we live with some things that we do not wholly understand. The earliest things were mechanical, but in the modern world we also have chemical, electrical, electronic, atomic, and soft things—things and systems of things of increasing complexity invented and designed to improve our daily lives and to help us achieve our aspirations as a sentient species. The goal is for the things we design to be realized, as much as is humanly possible, without failure.

Unfortunately, the potential for failure seems to lurk somewhere within almost every design. Murphy's Law states that if anything can go wrong, then it will. For technological systems, the law has the status somewhere between an absolute truth and a nervous joke. Citing Murphy's Law can lighten the mood at a design conference or during a critical test, but behind every smile and chuckle is a fear that this really is no joke. Indeed, if Murphy's Law is true, then it follows as a corollary that failure is inevitable. That would seem to be an overly pessimistic view, for the history of technology is full of examples of successful machines, structures, and systems. However, it is also the case that the successes were very likely preceded by instructive failures.

THREE

Designed to Fail

Failure in engineering, as in life, is understandably considered a negative occurrence—something to be avoided at almost any cost. When we hear that a device has broken or a system has failed, we begin to look for causes and culprits even as we are still searching for victims and cleaning up the wreckage. We want to identify how and why each failure occurs so that we can prevent a similar disaster in the future. Was the design at fault? Was the material inferior? Was the construction or manufacture careless? Were the components defective or flawed? Was the maintenance negligent? Was the use excessive? Was sabotage or terrorism involved? Such are the kinds of questions that are central to a failure analysis seeking answers. Such are the questions that must be asked and answered if we are to get to the heart of the matter so that, where appropriate, blame and liability can be assigned and, where appropriate, devices or systems can be redesigned and put back into service. We must also ask the appropriate questions to vindicate failed things and their designers when they are not at fault.

There is, however, another aspect to failure, one that from the beginning puts the usually pejorative connotation of the word in a much more positive light. We actually want certain things to break, for otherwise we would be frustrated in their use and possibly even harmed by their very existence. Sometimes, a component must fail for the larger system to succeed, or at least survive an insult to its integrity. The challenge to the engineer in this case is to design systems and devices that have well-defined and predictable failure modes and breaking points, so that such catastrophic physical phenomena as collapse or fracture happen in the way and at the time that they are supposed to. This is the case with the canvas roofs of outdoor stages. The roofs are designed with a weakly attached flap that is supposed to give way (that is, fail to hold) when the wind gets too strong, so that air pressure on the structure does not become greater than that which it can bear. In the summer of 2011 at the Indiana State Fair, when a stage's flap came loose before a concert it was apparently reattached in such a way that it held in winds approaching seventy miles per hour. That was well beyond the twenty-miles-per-hour wind speed at which it was designed to fail, and consequently the entire stage was blown over. Five people were killed and forty-five hurt because an intended failure did not occur.[1]

Some engineers object to associating the word *failure* with anything that works as designed, but if in fact something is designed to fail and it does so on cue, then the achievement should be acknowledged as simultaneously a failure and a success. It might seem ironic to call a failure a success, or a success a failure, but that is the way it sometimes is. Indeed, the Apollo 13 mission, whose planned landing on the Moon had to be aborted, has been described as a "successful failure" because despite an explosion that occurred when the spacecraft was two days along in its trajectory, the astronauts were able to salvage the system and return safely to Earth. The troubled flight was dramatized in the film *Apollo 13*, in which the actor Tom Hanks uttered the now-famous line, "Houston, we have a problem."[2]

The term "managed failure" has been used to describe situations where a certain kind of failure mode is designed into an otherwise robust system so that it is "capable of successful misuse." An automobile windshield, for example, can be expected to break under certain unintended but possible circumstances, which could include having a sizeable rock or an occupant's head strike it. The windshield might be made sufficiently strong to withstand the rock strike, but we would not wish it to be so strong that it is unyielding to a passenger's head during a collision. Designing the windshield to be shatterproof, so it does not fracture into dangerous shards of glass, is one way of managing failure through a sacrificial system. It is much preferable in the wake of an automobile accident to have to replace a broken (but not shattered) windshield than to deal with a fractured skull.[3]

As nature provides models for so many aspects of engineering, so it provides numerous examples of failure with desirable consequences. An eggshell is a wonderful structure for resisting the external pressures that an egg experiences during the process of being laid. Once an egg is delivered, however, the shell would be an instrument of extinction if it could not be easily broken from within by the beak of a weak inhabitant. Whether by cause or effect of evolution, the eggshell works. However, whether by design or coincidence, eggshells can also be broken almost as easily from the outside by an impact with the hard ground, by the pointed beaks and claws of predators, or by a sharp blow on the edge of a frying pan, thus providing nourishment for other species. The line between success and failure, between a good design and a bad, can be as thin as an intact eggshell or as fine as a barely visible crack in its side. Who has not inspected a dozen eggs in the supermarket, declared them sound, placed them gently into a shopping cart, made sure they were not thrown into the grocery bag, and avoided potholes on the way home, only to find a cracked egg or two upon transferring them from the carton to their place in the refrigerator? While this might be an annoying experience, it hardly qualifies as a disaster.

But missing a crack in an airplane fuselage, which engineers see as a shell of a different kind, can have catastrophic consequences. In 1988, this was the case with Aloha Airlines Flight 243, which was flying from Hilo to Honolulu, Hawaii. The Boeing 737 was at 24,000 feet when the fuselage ruptured explosively, leading to a decompression that swept a flight attendant out of the plane and peeled off a substantial portion of its roof. The catastrophic failure evidently began with a crack that had grown to critical proportions through a combination of corrosion and repeated loading and unloading—the phenomenon known as metal fatigue. One of the passengers reportedly noticed a crack in the fuselage when boarding the plane, but he did not mention it to anyone, perhaps thinking that inspectors must have thought it benign. In spite of missing a large portion of its skin, the plane did land safely on Maui, and the incident reminded the airline industry of the importance of detecting and reporting cracks.[4]

A similar incident occurred in the spring of 2011, but without any casualties. Southwest Airlines Flight 812 was en route from Phoenix to Sacramento when an explosive decompression occurred, forcing the plane to dive quickly to a lower altitude and make an emergency landing at Yuma. The five-foot-long tear in the fuselage was attributed to the existence of fatigue cracks along the entire length of the separated aluminum shell. The Boeing 737 aircraft had experienced about 40,000 cycles of takeoffs and landings, and cracks large enough to result in catastrophic separation of the plane's skin were not expected to begin to appear until about 60,000 cycles. Inspection of the failed section of fuselage identified manufacturing defects in the riveted joint as a most likely cause for the premature cracking and subsequent tearing of the structural skin. The rivet holes on the pieces joined were found to be "egg-shaped instead of round" and so did not align properly, resulting in higher concentrations of force than were anticipated by the airplane's designers. That neither the Aloha nor the Southwest incident resulted in the disintegration of the entire airplane and the loss of everyone onboard shows how far aircraft design had advanced since the mysterious failures in the 1950s of de Havilland Comets, the pioneering jetliners whose squarish windows led to the development of fatigue cracks that took almost everyone by surprise. It was the lessons learned from those failures that led to the outstanding successes of today's planes.[5]

In contrast with eggs, nuts are notorious for their hard shells, and some yield to being broken open for their fruit only under significant pressure or on the application of a mechanical device —a nutcracker—deliberately designed to cause cracking and failure of the shell. Indeed, we use the phrase "a tough nut to crack" to characterize some particularly difficult problem or puzzle that does not yield readily to a solution. Some nuts of unwieldy size, like the walnut, can be difficult to crack open even with the familiar spring-loaded hand nutcracker. But as many a nut nut with large enough hands has learned, squeezing two walnuts together causes one to yield to the hard-edged pressure of the other. Thus, two objects of the same design do not necessarily have the same resistance to failure. Which walnut succeeds to fail the other depends on a variety of extra-design circumstances, including how the nuts are arranged in the hand and which nut might have had its shell slightly cracked open in an earlier elimination round of nut-to-nut combat.

Physical civilization as we know it has long depended on the existence of breaking points and failure planes in nature. The forming and shaping of stone fragments through knapping, wedging, and thermal fracturing gave us our earliest tools, weapons, and stones for constructing monuments and other structures. The friability of soil and the brittleness of rock allowed early agriculture and mining to develop. Everything in nature has its breaking point, thank goodness; the hardest diamond yields to being cleaved and, we think, enhanced. The atom, as its very name tells us, was once thought to be indivisible. With the splitting, or breaking up, of the atom, a new source of energy was released for the use and, in most cases, the benefit of humankind. Can anyone today say with more certainty than Democritus what will be the ultimate understanding and use, or abuse, of subatomic particles?

Animate nature also is known to make good use of failure. Tropical spiders of the genus *Pasilobus* make horizontal webs of a triangular design, with sticky threads drooping between radial lines. When a moth or other insect brushes against one of the sticky threads, it breaks near one of its two ends and leaves the snared prey dangling below the web proper. Since it is more difficult to tear loose from a dangling thread than from a drooping thread, the moth is unlikely to escape. Alerted by the failed thread, the spider need only move to that location and haul in the line and its attached catch.[6]

The success of our human-made environment, which consists largely of the collective products of engineering design, is possible because of the controlled fracture, breaking, crushing, and general triggering of desirable failure modes in raw materials, allowing them to be refined and reshaped and recombined into useful things. And each of those useful things, because of the very nature of its constitution or reconstitution, has its own breaking points and preferred failure modes. Think of trees, which are first felled by ax or saw or wind, then cut up into firewood or lumber, and ultimately used to feed a fire or to make tools, furniture, or shelter. The challenge to engineers throughout the ages has been to understand how and why everything, both natural and artifactual, fails and how to exploit that knowledge and experience in devising things that do not fail under normal intended use—unless the failure itself is intended, as it sometimes is.

The boundary between success and failure—between intended and abusive use—can be as narrow as the edge of a sharp knife. A paring knife, for example, can be effectively used to peel an apple, for that is among its intended purposes—to cause the failure of the skin to continue to adhere to the fleshy part of the fruit. The knife could also be used to peel an orange and even slice a carrot. But it might not be as effective in peeling tougher foods, such as a coconut, for then the implement itself would soon likely fail to be useful, if not by breaking then by becoming dull. But a dull knife can be sharpened by drawing it at a shallow angle across a whetstone. However, if the knife blade were to be held perpendicular to that same whetstone and moved in a sawing motion, the blade's edge would likely soon be ruined. Other deliberate misuses of the knife, such as employing it as a tool to pry the lid off a paint can, could result in a bent or broken blade, the latter failure threatening to harm whomever might be in the vicinity of the scandalous act. There is a proliferation of specialized tools and artifacts in large part because of the narrow window of safe and intended successful application for each of them. Used outside that window, they can fail.

The ongoing design and redesign of tools and related devices can be viewed as the refinement of how they cause controlled failure in the things on which they work. It can also be viewed as a reaction to how the tools fail to function properly and efficiently. Power tools have widely displaced their hand-operated counterparts because, generally speaking and under the control of a skilled operator, they make it possible to achieve a desired objective more easily, quickly, and neatly than the equivalent hand tool. And those objectives are usually to effect failure in the materials and parts being worked on. In carpentry, for example, the action of the circular saw interrupts the continuity of wood fibers and thus causes a piece of lumber to fail to adhere to itself; the screwdriver, in conjunction with the screw, causes wood to fail to resist the connecting device, as does the hammer in conjunction with the nail.

Even among some of the most commonplace and ordinary things, the idea of deliberate and controlled failure is essential to effective use. The first adhesive postage stamp was introduced in England in 1840. Known as the Penny Black, the one-penny stamp bore a portrait of Queen Victoria but no more explicit indication of the country of issue. (In recognition of its innovation in pre-paid postal service, Great Britain remains to this day the only country in the world allowed by the Universal Postal Union to is-sue stamps bearing the image of its reigning monarch in lieu of the country's name.) The Penny Black and other early postage stamps were printed in multiples on sheets without perforations. This meant that to separate the individual stamps, the user had to cut the sheet with a pair of scissors, an action accompanied by the danger of failing to cut in a straight line and thereby ruining a bor-dering stamp. Perforated sheets of stamps were a distinct improve-ment, providing as they did a preferred line of separation of con-trolled and desirable failure. Interestingly, the design of modern stamps, which come with a self-adhesive that does not have to be licked or otherwise moistened, often incorporates serrated edges that no longer serve the historical purpose. So evocative of the stamp is the ragged-edge feature that sheets or strips of modern stamps contain wavy laser-scored lines that allow one stamp to be separated easily and neatly from its neighbor along the faux-serrated pattern.

Perforations are, of course, encountered in many different con-texts, and they were once readily introduced as part of the printing process. It was easy to incorporate into the letterpress type set up to print a card or coupon, for example, a dotted-line rule (typo-graphical jargon for a linear piece of type) that would incise the paper at the same time as the text was being printed upon it. For the advertiser not to have specified perforations would have been to risk readers not using an order form when scissors were not handy or because, in attempting to separate the form by creasing and tearing, the form itself became damaged and unusable. Perfo-rations ensured that a sequence of small rips (material failures) occurred easily and along predetermined lines so that the customer had little excuse not to use an order form or response card just because it was not convenient. And imagine a checkbook without built-in failure lines.

With the advent of offset printing, in which the paper is not physically impressed by type or rules, perforations were not so easily produced and generally required a separate operation. As a result, the printed (but not incised) dotted line, sometimes with a cartoon pair of tiny scissors indicating that real ones were to be used, came to indicate the boundaries of a coupon or reply slip. This meant, of course, that readers of magazines, newspapers, and advertising inserts had preferably to have scissors nearby, something that was not always convenient. Such conditions are what lead to the invention of new devices for clipping coupons and other items, and the newspaper clipping device sold as a Clip-it became a marketing success. This small curved blade, encased in a plastic holder shaped like the foot of a sewing machine, could be navigated around coupons, news items, features—virtually anything one wished to remove from the newspaper to save or send to a friend. Subscription reply cards are still inserted in newspapers and magazines, but they do not always come with visible perforations or tiny scissor icons. Tear lines are now scored by laser, and the edge that results from tearing along an electronically formed line can appear to be as sharp as one produced by a paper cutter.

Sometimes we find ourselves wishing to remove an engaging newspaper story or hilarious magazine cartoon that is not framed by perforations or laser lines. The producers of such publications, which are printed on both sides of the paper, could hardly provide preferred tear lines around each item of potential interest, and so we are left to our own devices. As a rule, we will have no scissors or Clip-it in our pocket or purse, and trying to tear a sheet of newspaper across its grain invariably leads to the rift curving into the grain and thus into the coveted item. Failures, once begun, tend to follow their own course. Most people avoid this problem by neatly folding the paper along the line of an edge they wish to create, creasing it decisively, and then folding it back upon itself before attempting a tear. Repeating these actions a couple of times before commencing the tear will generally result in a rather neat if fuzzy edge. What we are doing, of course, is breaking the paper fibers that lie across the desired separation line. Folding the paper back and forth is akin to bending a metal paper clip back and forth. In other words, we are effecting what might be termed a successful fatigue failure. The paper itself is not designed to fail, but by our actions we have redesigned a selected section of it to do so.

We do not have to work so hard to get at other things that we desire. A variety of food products, such as candy brittles, chocolate bars and barks, and crackers, are prepared in sheets larger in size than is convenient for consumption. Tradition generally dictates whether the broken sheets are expected to have regular shapes. Pieces of peanut and other brittles have characteristically irregular shapes, which can be created by smashing the hardened candy on a harder surface, where it breaks like a sheet of glass. The more controlled breaking of actual glass is, of course, achieved by scoring or scratching the surface to introduce a line of weakness defining a preferred size or shape. In a similar fashion, hard candy or cough drops once were made in sheets that had thinner and thereby weakening lines pressed into the sheet during the forming process and so broke into predetermined shapes when thrown into a packaging hopper or candy jar. Softer chocolate bars likewise have thinner sections between the bite-size pieces, and these serve for less violent separation into portions.

Crackers, ranging from saltines to matzos, often come in perforated sheets of two or four or more portions. Producing the crackers in attached multiples simplifies and makes more efficient the baking and packaging process, but customers might be discouraged from buying a particular brand again if its parts failed to separate cleanly and neatly, especially for hors d'oeuvres served before an important dinner. Just as we do not expect unintended failures to occur in our machines and structures and systems, so we do expect deliberate and designed-in failures to occur in just the intended way. But achieving a clean break line often takes more luck than skill. Some rustic crackers come in large sizes but are not perforated, and these present dilemmas for hosts wishing to serve neat sections of them with cheese. They might not fail to please the palate, but they could fail to please the eye.

Not infrequently, the failure-constrained design problem is to contain something securely in a package that does not open accidentally during shipping or handling but that does open easily when the end user wants to get at the contents. The solutions to such bipolar design problems can produce some of the most satisfying and some of the most frustrating things to use. Although its development took decades from Whitcomb Judson's 1893 patent for a "clasp locker or unlocker for shoes," the "perfected" zipper is a classic example of something that makes easy—indeed, almost enjoyable—the work of the closing and opening of countless things. Invented in the mid-twentieth century, the plastic bag incorporating a toothless zipper has become ubiquitous—especially at airport security checkpoints—but some versions tend to be more difficult to close successfully than to open.[7]

That is not the case with all plastic or foil bags. The little sacks of peanuts that used to be distributed freely on airplanes are a notorious example of frustrating packaging. In the best of cases, sealed snack bags certainly hold and protect their contents in a puffy airtight foil pillow, but many a hungry would-be snacker has experienced frustration in trying to find the intended failure-initiating trick to get to the snack. The most aggressive pulling, twisting, and tearing at the location marked "tear here to open" can all too often result in nothing but a disfigured but still intact bag—or an explosive snack spill. The determined peanut or potato chip lover may have to resort to using a knife, scissors, nail clipper, or teeth to attack the package. When a packaging designer's criterion for desirable and controlled failure is overly biased toward keeping the contents safe from predators, intended consumers have to become designers themselves. But packaging that does not open easily is as poorly designed as packaging that does not hold its contents securely. Prescription medicines have for some time now come in child-resistant and tamper-proof packaging that is adult-resistant too. And who has not purchased a tool enclosed in a plastic shell whose opening seemed to require the use of the tool itself?

Many a parent and child have been frustrated on Christmas morning by their inability to free a colorful toy from its plastic or cardboard cage. Modern packaging has become so focused on display and security that springing the prize from its prison can seem to take the tools, skill, and cunning of a safecracker. In defense of package designers, it is no easy task to come up with something that protects a product during shipping, allows for its attractive display on a store shelf, keeps it secure and intact and fresh looking until bought, and can be opened easily by the purchaser. This collection of objectives, like those of virtually any problem in design, is inherently self-contradictory. It is no trivial matter to foil prying fingers in the store and yet welcome them at home.

Making bottles air- and spill-tight has been the pursuit of inventors seemingly as long as there have been bottles—and not just of wine. At the end of the nineteenth century, a spate of patents was granted for capping bottles of liquids of all kinds. The bottle cap with cork insert emerged as the dominant design (in conjunction with the lipped bottle, of course), and it relied for its application on the crimping of metal—a form of intended controlled failure—and for its removal the prying off of metal—another form of failure. Of course, in the time between the application and removal the cap was not supposed to fail to hold tight, even under pressure to pop. That primary function at first received more attention than did that of a controlled failure mode that allowed access to a bottle's contents. The bottle opener was designed to facilitate failure when it is desired. Unfortunately, bottles and bottle openers are not always found in the same place at the same time.

Thirst can be a mother of improvisation, however, and some of the clever (and questionable) ways to cope have been recorded in the book *99 Ways to Open a Beer Bottle without a Bottle Opener*. But while using the spokes of a bicycle wheel or the hinge on a door to pry off a cap may work, it can also cause the loss of some of the bottle's sought-after contents. Using anything not designed for the task at hand can end in more than spilled beer. The introduction to the 99-ways book does acknowledge that bottles fitted with twist-off caps do not require a mechanical opener, but that does not mean that we do not sometimes think we have to search for one. Indeed, it often does not occur to beer drinkers who favor imported brands to try to twist a cap off an unfamiliar domestic brew; they use a bottle opener to remove even a twist-off cap, often breaking the neck of the bottle in the process. Conversely, beer drinkers who are accustomed to brands that employ twist-off caps instinctively try to twist off any bottle cap. All of us tend to make assumptions about how something is designed, how it works, and how it is supposed to fail or not fail, and we proceed accordingly.[8]

No matter how they are closed and opened, glass bottles have their own undesirable failure modes—including that of shattering when dropped—and so the tin can came to be the container of choice under certain circumstances. But the frustration of being on a picnic without a can opener spurred the inventor Ermal Fraze to come up with the pop-top for the beverage can, a model of a designed-in failure feature. The first pop-top tabs separated completely from the can, which threatened the very existence of the opening device. Only the development of can tops that retained the metal tab kept pop-top can–banning legislation from being implemented in the environmentally sensitized 1970s. But the pull ring on many of the tab-retaining cans sits so close to the top itself that stubby fingers cannot easily get beneath it, and it threatens to break any fingernail that tries to lift it. This problem has led to the development of keychain devices that give the proper leverage without using the fingernail. Perhaps the most unforgivable failure of all is the failure to anticipate the downside of an otherwise good design.[9]

We also rely on designed-in failure modes to keep us safe; sometimes a component must fail for the larger system of which it is a part to survive an insult to its integrity. The controlled failure of a fuse protects us from an overloaded electrical circuit that could overheat and cause a fire. Sprinkler systems are designed to be activated (by their failure to contain the water they contain) when fires do occur, and the key device in many of these systems is a thermomechanical fuse that softens in the heat of a fire and thus relaxes its push against a valve in the pressurized piping. Pressure-relief valves on water heaters are another kind of fuse, one designed to fail at a predetermined pressure, one above that at which the heater is designed to operate safely but below that which would cause the heater itself to explode.

Fuses of a different kind are an integral part of the design of the new east span of the San Francisco–Oakland Bay Bridge, which is expected to be completed in 2013. The bridge is, of course, located in an area prone to earthquakes, and indeed it was the damage done to the bridge by the 1989 Loma Prieta quake—and the weaknesses in the structure that it revealed—that led to the two-decades-long project of replacing the half-century-old structure. The centerpiece of the new span is a distinctive self-anchored suspension bridge that has only a single tower, which consists of four closely spaced steel pylons connected by steel struts, known as "shear link beams," which might be likened to a chain's weak links. To ensure that some future earthquake will not do irreparable damage to this signature structure, the tower struts will act as structural fuses to absorb the quake's energy. While the sacrificial fuses might be permanently damaged in the process, they will keep the main components of the tower proper—the pylons—from suffering the same fate. The bridge's long concrete approach viaduct has also been designed with steel fuses known as "hinge pipe beams" located inside the structure, between critical sections of the span. These will also be permanently deformed by a large enough quake, but they and their counterparts in the tower will be able to be removed and replaced relatively quickly and easily.[10]

Many a technological system contains a weak and sacrificial part. Fire-alarm boxes are often found faced with glass, no doubt to discourage impulsive and inadvertent false alarms. Unfortunately, the little glass panels on fire-alarm boxes may not fail easily in the heat of excitement, and precious time could be lost in trying to break the glass with a soft fist—yet if hardened to success, the fist could be cut and bloodied. To avoid such bodily harm, glass-enclosed fire-alarm boxes are equipped with little hammers, so that the glass can be broken as quickly and cleanly as possible. It takes a deliberate act of benign vandalism, that is, breaking the little glass window, to turn in an alarm.

We often encourage one mode of failure to obviate a less desirable one. Concrete sidewalks or driveways tend to crack as they cure because the cooling mass of concrete wants to contract, but the length or expanse of the pavement also wants to hold its place. Thus the surface is finished by carving into it well-spaced grooves or aesthetic "cracks"—those lines upon which children are not supposed to step, lest they break their mother's back. These grooves act not unlike perforations and provide preferred sections along which the curing (and shrinking) concrete can physically crack or tear to relieve the internal stresses that build up in it. Since the physical cracks form at the bottom of an aesthetic "crack" or groove, they are hardly noticeable and hence the appearance of the walk or driveway is not marred by disfiguring random, jagged, and uncontrolled cracks. Any failures of the concrete to adhere to itself have been corralled.

The concrete in massive structures also wants to crack as it cures. Such cracks could, of course, result in more than cosmetic flaws; they could result in a leaking dam. Hoover Dam is famous for having had its concrete poured in large blocks with cooling pipes embedded in them to carry away the so-called heat of hydration in a controlled manner. When the concrete was sufficiently cooled, the gaps between the shrunken blocks were filled with cement grout effectively to produce a monolithic structure that was further compressed together by the water it impounded.[11]

Our highways have become another site of designed-in failure modes intended to ameliorate more harmful failures. The hoods, fenders, frames, and other components of modern automobiles are designed to behave in a crash by crumpling in a most benign, if not inexpensive, energy-absorbing manner. Introducing crumple zones and other locations of preferred failure in automobile body components was pioneered by manufacturers like Volvo. The crashworthiness of such cars directs as much as possible any unwanted energy into causing failures outside of the passenger compartment. And well-designed guardrails beside the road are intended to fail sacrificially under the impact of a speeding vehicle and absorb energy that might otherwise go into injuring the vehicle's occupants. The ends of guardrails present an especially difficult design problem, and the challenge is to get them to fail in an energy-absorbing mode rather than function in a piercing and penetrating mode that causes harmful consequences in a vehicle impacting them. Until such friendly failing end fixtures were devised, the familiar clusters of yellow plastic drums full of sand and water were placed between the end of a guardrail and oncoming traffic. The sacrificial failure of the drums, which are still employed in many places, is designed to slow an errant vehicle to a reasonably safe stop.

In some cases, things are designed so conscientiously against failure that they present challenges when they have outlived their usefulness. Nuclear power plants, because of their potential to wreak so much havoc should an accidental release of radiation occur, have been built with extreme strength—in some cases enough to withstand the impact of an airplane. This resistance to failure, coupled with the residual radioactivity in decommissioned plants, has presented engineers and society with a new challenge: how to disassemble a plant that is no longer needed or wanted. But for all their thick reinforced-concrete walls, nuclear containment structures do have their failure points that will yield to properly designed and chosen wrecking tools, even if they have to be wielded by robots.

Other large structures, such as tall buildings, long-span bridges, and massive sports stadiums can also reach or exceed their useful life and so need to be removed, either before or after a replacement is available. In skyscraper cities like New York, where land is scarcer than clean air, it is not uncommon for a perfectly serviceable structure of modest height to be demolished to make way for one of more stories. With sports stadiums in the age of television, it is not necessarily so much the need for a larger capacity as for more amenities, such as luxury skyboxes, that can motivate the destruction of a once-hallowed venue. Sometimes, safety is also an issue.

This was the case with the revered Yankee Stadium, "the house that Ruth built." The original structure, which dated from 1923, was renovated in 1976, but in the 1980s there were calls for a new ballpark, with the Yankees' outspoken owner George Steinbrenner claiming that the existing one was unsafe. Indeed, in 1998, just days short of the stadium's seventy-fifth anniversary, a 500-pound piece of the structure fell without warning from beneath the third deck onto seats on the second. Fortunately, there was no game being played at the time, and so no one was hurt in the incident, but it did emphasize the deteriorating condition of the aging structure. After much posturing and political maneuvering to keep the team playing in New York City, construction of a new stadium began in 2006. The construction site was located on parkland just across the street from the old ballpark, in which the Yankees continued to play games. In 2009, with the completion of the new stadium, which was designed deliberately to resemble the old, demolition of the old one could commence.[12]

With the new stadium and the transportation infrastructure that effectively brought fans right up to the entrance of the old ballpark so close by, demolition had to proceed in a tightly controlled and low-risk manner, lest collateral damage be done. Heavy equipment staged on the famous ball field was used to tear down large sections of the structure at a time, pulling them inward and away from the new stadium and the elevated train station. Area residents, waiting for the site to be cleared so that they could regain a neighborhood park, became impatient with the slow progress of the job, which was not completed until 2010. The use of explosives to implode the old stadium—as was done in 2000 to bring down the Seattle Kingdome—would certainly have speeded up the process, but at considerable risk to the surrounding area. Fortunately, in many such cases risk is sufficiently low to allow the more dramatic—and usually quicker—means of demolition to occur.[13]

Obsolescence and safety considerations can also demand the demolition of an old bridge that is too narrow or too weak for the expectations and demands of today's traffic. It is not that the bridge was improperly designed for its time; it is just that design criteria have evolved to meet new needs or that the old design is obsolete in strength or size for newer times and vehicles. But disassembling a sizeable steel bridge is no easy or risk-free task, and it can be a time-consuming and expensive undertaking. Thus, a common way to achieve the desired end—to remove the bridge in the quickest, safest, and most economical way—is often to cause it deliberately to fail and drop into the water below, from where the pieces can be dredged up and sold as scrap. A typical means for achieving such an end is to plant explosives at critical points on the structure and bring it down in a dramatic fashion.

This was the plan for removing two obsolete cantilever-truss crossings—the 1929 John P. Grace Memorial Bridge and the 1966 Silas N. Pearman Bridge—with which the citizens of Charleston, South Carolina, had developed a love-hate relationship over the years. In 2005, after traffic was transferred to the new cable-stayed bridge—the Arthur Ravenel Jr. Bridge—that had been built beside them, the main spans of the old Cooper River bridges were to be fitted with explosive charges that when detonated would cause failures to occur at critical locations and so bring down the structure in manageable pieces. Establishing the proper location, size, and sequence of detonation of each of the charges to cause controlled failure was as much of an engineering problem as was designing and building the bridges not to fail in the first place, and calculations and drawings signed and sealed by a professional engineer had to be prepared. The demolition process began with the removal of the signs, lights, and other appurtenances on each bridge's roadway, followed by the removal of the roadway itself. Beams spanning over Drum Island were severed by machines with massive jaws and were allowed to drop onto the unoccupied island, from which they were removed. Concrete beams over populated areas and recyclable steel girders were carefully removed by cranes, which placed the salvaged parts on trucks. Wherever possible, high concrete piers were brought down by dynamite and then broken up on the ground. Explosives were also used to break up support structures located over and under water.[14]

To deconstruct the main spans of the bridges—the cantilever trusses over the navigation channels—workers first removed their concrete decks and then the stringers between the girders. This left a pair of lacey steel superstructures on which shaped charges were strategically placed so that the blasts would sever the skeleton spans into twenty-to-forty-ton sections that could be retrieved from the water. To make that task easier, the parts were rigged with retrieval cables and buoys before the explosives were set off. The first spans to be blasted into the water were those over Town Creek, which contained the side channel of the river. This provided an opportunity to test the scheme before repeating it over the river's main shipping channel. Surprises there were out of the question, because the penalty for not opening that channel on time was $15,000 per hour, or $360,000 per day late. It was fortunate that there was a practice try, because not everything went smoothly: the planned failure scheme failed to go as planned. One large portion of the superstructure was left intact, and it took three weeks to recover all the steel out of Town Creek. An improved method used on the creek's second bridge, which left it in larger pieces, enabled the retrieval process to go more quickly. The disappointing results with explosives led to an entirely different approach over the main channel of the Cooper River. Instead of blasting the bridge apart and letting the parts fall into the water, workers cut up the bridge in place and lowered the parts onto barges. With parts as heavy as 600 tons, special heavy lifting and lowering equipment mounted on barges had to be used, resulting in what was a more expensive but also a more reliable operation. The steel trusswork was barged upriver to be cut apart and recycled. Concrete rubble was barged out to the ocean, where it was dumped in preapproved locations to create artificial reefs. Not all demolition jobs have such problems or complications.[15]

The eighty-year-old Crown Point Bridge was a vital link across Lake Champlain, connecting Fort Crown Point, New York, and

Chimney Point, Vermont. As early as 2006, plans were announced to address the deteriorating condition of the bridge. In 2009, a routine inspection of the structure detected much larger cracks in its concrete piers than had been found in the prior biannual inspection, indicating a condition of accelerating deterioration subsequently confirmed by divers conducting an emergency underwater inspection. The concrete piers, which were designed and constructed to 1920s standards, were not reinforced with steel, as would be the practice today. Without reinforcing, the piers were less able to resist the push of ice, which tends to form first near the shore, where the water is shallower than in the center of the lake, and so pushes unevenly on the piers and tends to bend them. Such a scenario was evidently not anticipated by the original design engineers, and it was not made manifest before the unreinforced piers began to show signs of increasingly bad cracking. The installation of devices to monitor the movement of the piers confirmed that they were effectively beyond repair.[16]

Since the piers were so weakened by the cracks, the bridge itself was believed to be in danger of falling into the lake. Within weeks of the emergency inspection, a report was filed and the bridge was closed to traffic. Not only was the spontaneous breakup of its supporting piers a possibility, but also the aging steel superstructure had developed frozen bearings, a common ailment akin to the arthritic conditions that aging humans experience in their joints. The closing of the Crown Point Bridge created considerable hardship for those who had relied on it to commute to their jobs across the lake, for without the bridge they had to drive an extra hundred miles each way. To alleviate the inconvenience somewhat, an emergency project was undertaken to construct approach roads, parking lots, and other facilities to institute temporary ferry service until a new bridge could be completed. That was expected to occur by 2012.[17]

A weak or faulty bridge naturally presents considerable risk to those who use it, and that risk can be completely eliminated by closing the bridge. But in the case of the bridge across Lake Champlain or over any recreational waterway, there is also the risk of a deteriorating structure falling on boaters. Since boat traffic moving under the thousands of feet of steel trusswork of the Crown Point Bridge would be more difficult to control than the two lanes of road traffic that had moved over and through it, the decision was made to demolish the bridge as soon as possible—preferably before the weather warmed up and the boats came out in force. In December 2009, the old bridge was brought down with the detonation of five hundred charges packed with a total of eight hundred pounds of explosives. Such instant demolitions often make a big splash on the television evening news, and there is no shortage of YouTube videos of them on the Web. The debris that fell into Lake Champlain was to be removed by the spring, when construction of the new bridge was to commence, with a completion date set for 2011. The new bridge, which was variously described as a "modified network tied arch bridge" and a "steel structure with a handle-like arch along the main span," was designed to have a life expectancy of seventy-five years, which is remarkably close to the age to which the old span survived.[18]

Sometimes, as with the Cooper River bridges, a planned failure of a land-based structure can be a failure itself. Old, tall smokestacks, which may once have been landmarks, are often the object of demolition to clear the way for something new. Setting off a well-placed explosion or series of explosions at the base of a tall, slender structure is usually sufficient to topple it. When done correctly, the explosives will bring a stack down as effectively and precisely as a properly wielded power saw can a tree. But whereas a sawn tree can be expected to come down with its trunk intact, a falling brick chimney will often break apart before it hits the ground. This happens because while the brick structure may act like a monolith as long as it is standing up straight, when it is falling the forces induced by the accelerating mass of the structure become greater than the mortar can bear. The chimney thus breaks apart at the point at which these forces first overcome the adhesive strength of the mortar. Exactly where that place is depends upon the shape and taper of the tower of bricks, and also upon how tenaciously the mortar holds the bricks together. For a typically shaped chimney with deteriorated mortar, the bricks will begin to separate close to halfway up the stack very soon after it begins to topple; for a chimney in which the mortar is strong, the break will more likely occur later in the fall and about one-third of the way up from the bottom.[19]

Some tall chimneys are made of reinforced concrete, in which case the acceleration forces generally cannot become large enough to cause the concrete to separate during the structure's fall. This is exactly why reinforcing steel is embedded in the concrete in the first place; it helps keep a chimney from cracking in the wind. When such a chimney is demolished, it will fall in one piece, perhaps breaking up somewhat only when it strikes the ground. It is very important, of course, for the demolition explosives to act on the base in just the right way so that the chimney falls in the proper direction. This did not happen at the ninety-year-old Mad River power plant in Springfield, Ohio, where in the fall of 2010 a 275-foot-tall concrete chimney was being demolished. Instead of falling to the east, as planned, the chimney fell to the southeast, severing two high-voltage power lines and damaging a pair of turbines at the plant, causing a power outage for about four thousand customers. Speculation was that undetected cracks existed in the chimney and were not taken into account when the demolition crew placed the seventeen pounds of explosives that were used. According to the "chief blaster," it was only the fourth time in his thirty-one years of imploding structures that the job did not go as planned. In fact, it was only the second time this had happened in his last twenty-five years of work. As evidence of his expertise, he explained, "I just successfully 'shot' an 85-ft-tall building in Athens, Greece, with only 15 ft of clearance on two sides and 30 ft of clearance on the other two." That's the way demolition is supposed to work—and usually does.[20]

Old or defective buildings of all kinds are frequently the object of designed failures. The 376-foot-tall Ocean Tower high-rise condominium on South Padre Island, Texas, was never occupied, because the structure began to lean even before it was completed. Uneven ground settlement was blamed for the problem that the uncommonly tall structure experienced on the sandy beachfront location, and the owner of the building decided to bring it down. He contracted the company Controlled Demolition to implode the building, causing it to fall in upon itself, without damaging nearby houses, sand dunes, and a park. The demolition firm described the task, which would set a height record for an imploded reinforced-concrete building, as "one of its most challenging razings using explosives." In fact, the company considered the job to be "its second most difficult implosion after Seattle's Kingdome," the stadium that in 2000 was brought down by over two tons of dynamite.[21]

The Texas island job was complicated for a variety of reasons. First, the unusual height and slenderness of the structure did not provide much opportunity to force its sides inward before the tower fell in the clutches of gravity. Further complications revolved around the fact that the three-sided condominium tower was built on top of a rectangular multistory parking garage, whose structural beam system included a weak section to allow for passage of the base of the tower crane used during construction. Additionally, the condominium was founded on sand, which is a good conductor of vibrations that can do damage to nearby structures. In order to deal with all these complications and at the same time ensure that the falling structure and debris would be contained within a limited area and would not rain down on neighboring structures or onlookers, Controlled Demolition needed to design and plan carefully. Fortunately, the 2009 demolition of Ocean Tower proved to be a model of thoughtful design, planning, and execution of a controlled failure. Strategically placed charges, backed up by debris-containing screens made of geotextiles— mat-like materials used to confine soil and rock—were set off in a precisely timed sequence that in the span of 12.5 seconds straightened the tower, leaned it about fifteen degrees toward the Gulf of Mexico, and then let it drop vertically into a well-contained pile of broken concrete and twisted steel. It was indeed a successful failure.[22]

As the 1993 explosion in the basement public parking garage of the north tower of New York's World Trade Center demonstrated, just setting off a truck full of explosives may not be enough to cause a building to fail. The truck bomb did do a lot of damage, taking out multiple basement floors and leaving a gaping hole behind, but the building itself did not come down the way the terrorists evidently had hoped it would. To bring down a building in a controlled failure mode takes a lot more than just a truck full of dynamite or other explosives. It takes considerable preparatory work and strategically placed charges that are detonated in the proper sequence. In other words, to cause a structure to fail in a certain way is as much of a design problem as was ensuring that the building stood up properly in the first place.

In 2001, the collapse of the World Trade Center towers in New York within an hour or two of their being struck by hijacked airliners looked to some observers very much like controlled demolitions. Indeed, conspiracy theorists attributed the cause of the collapses to the detonation of well-placed explosive charges, offering as supporting evidence what were described as puffs of smoke that were emitted during the pancaking of the towers' floors. The collapses were certainly reminiscent of controlled building demolitions, but that does not mean that that was what they were. On the contrary, the way the twin towers collapsed progressively under the weight of their own falling floors was evidence of why building demolitions work so successfully. A falling building effectively demolishes itself, and so the design problem faced by a failure engineer—if that term may be allowed—is to initiate the collapse in the most efficient and effective way and then let gravity do the heavy work.

The design of a demolition involves more than just establishing the size and location and firing sequence of explosive charges. A lot of the design involves preliminary work in removing or weakening strategic columns and other structural impediments to progressive collapse. Some columns are cut partially or completely through to ensure that they will be moved sideways (and so out of the way of falling floors) by the action of the explosion. Support walls are removed so that they will not offer any resistance to the collapse. The building, in other words, is reduced to a weakened structural frame before any explosives are even put in place. In the case of the September 11 attacks, the fires that were ignited by burning jet fuel and fed by office furniture and supplies did the work of weakening the structural columns that had not been taken out by the plane impacts. The temperature of the flames naturally heated up the steel columns, which in time became softened to the point where they could not support the load that bore down on them. As a column gave out, its weakened neighbors were immediately required to take on additional loads, but because they too had been heated and softened they could not do so. Once the columns on the impact-damaged floors had reached the limits of their strength to hold up the floors above, those floors fell with a force greater than the columns on the floor below—also weakened by the heat of the fire—could bear, and so the buildings collapsed progressively down to the ground. As floor fell onto floor with increasingly greater force, considerable impact pressures crushed anything between them, and dust and fine debris were pushed outward by the rapidly compressed air. No explosives were needed, because the heat of the fires weakened critical support columns and the momentum of the falling floors provided sufficient energy to pulverize concrete and push steel columns aside.

Whether the terrorists who planned and executed the 2001 attacks on the World Trade Center had designed them as carefully as a professional demolition team would have may never be known for sure. What does seem likely, however, is that the terrorists did learn from prior failures. The failure of the 1993 basement truck-bomb explosion to cause the north tower to topple into the south tower taught the evildoers that another strategy was in order. The terrorists may well have designed their airplane attacks to bring down the buildings, but they might not have known exactly where to impact them for the maximum likelihood of success. Seeing that the crash of the first plane did not immediately cause the building hit to fail catastrophically, the hijacker piloting the second may have reasoned that he should direct its impact at a lower floor, so that there would be a greater portion of the building's weight bearing down on the section with the severed columns, which numbered many more than the truck bomb had damaged. That such a decision would be a diabolically effective one was demonstrated by the fact that the tower struck second was the first to fall.

Failure, in short, can be by design, whether for good or ill. It can be an end unwelcome—and possibly unimagined—by anyone but terrorists. But failure can also be a welcome and beneficial end, whether in providing the makings of an omelet or disposing of an abandoned or unwanted structure. We rely on desirable failures of all kinds. They are designed into many of the products we use every day, and we have come to depend upon things failing at the right time to protect our health and safety. We also count on the inevitable failure point of materials and structures to make technological developments reversible. Failure is a relative concept, and we encounter it daily in more frequent and broad-ranging ways than is generally acknowledged. And that is a good thing, for certain types of desirable failures, those designed to happen, are ones that engineers want to succeed in effecting.

In general, though, failure in engineering is something to be avoided, and the goal of engineering design is to make sure it does not happen when or where it is not wanted. Thus a structure like a bridge or a building is supposed to be strong enough to withstand whatever anticipated natural or manmade challenges are thrown at it, whether they be heavy snows, high winds, strong earthquakes, massive floods, excessive overcrowding, or terrorist attacks. Not every challenge will be as likely to occur in every location, of course, but earthquakes cannot be ignored in California or hurricane winds on the Gulf Coast. It is the engineer's responsibility to understand what the relevant and likely challenges are in each particular case and to design structures accordingly.

However, while structures can be designed to resist every possible and imaginable onslaught, judgments have to be made about whether that is practical and affordable. Natural hazards are, by nature, not fully predictable. A city like New Orleans is clearly threatened by hurricanes, and it is virtually certain that another big one will strike within, say, the next century. The same goes for San Francisco and earthquakes. But exactly when the next consequential event will occur and how strong it will be are more matters of chance than of precise scientific prediction. Engineers charged with designing levees and buildings and bridges and other structures thus have to deal in probabilities rather than in certainties. They must calculate risks and benefits and consequences and present the results to government bodies or boards which can authorize the expenditure of enormous amounts of money to design and build structures that will not fail. But under what assumptions do such calculations and judgments take place? They invariably have to take into account and weigh, implicitly or explicitly, the cost and affordability of fail-safe improvements against the value of human life.

How much risk is considered acceptable is highly dependent upon political and cultural factors. The Netherlands has a long history of defending its land against the sea, and its dikes and other defenses are legendary. Still, after the massive flooding that followed a 1953 storm surge, the country undertook a thirty-year plan to prevent that from happening again, or at least to reduce it to an acceptably low level of risk in the case of a storm that could be expected to occur only once every 10,000 years. Decisions about what to protect and to what degree were based on risk analysis that took into account not only the probability of a breach of defenses but also the projected cost of damage that would result from the failure, which included not only that of rebuilding but also that of economic impact on the nation.[23]

With new issues like global warming and sea-level rise high in their consciousness, the Dutch acknowledged more recently that their existing level of protection against massive flooding may not have been sufficient. In 2007, the country's parliament charged its expert-filled Delta Committee with determining what it would take to "climate-proof our country for the next 200 years." Whereas a 1 in 10,000 risk had been considered acceptable, now a 1 in 100,000 risk was considered desirable, at least for such economically critical areas as surround Rotterdam, Europe's largest port and the source of 65 percent of Holland's gross domestic product. To lower the risk to that level for all areas of the country would be prohibitively expensive, however, and so it was deemed necessary to protect rural sections only to a level of 1 in 1,250 or even lower. Such value-laden decisions may not be politically achievable so easily in America, where only in the wake of Hurricane Katrina did Congress direct the Corps of Engineers to upgrade the defenses of New Orleans to the point where risk of flooding was 1 in 100. According to statisticians, this means that the chance of a hundred-year flood occurring during the lifetime of a thirty-year home mortgage would exceed 25 percent. Acceptable levels of failure, cost, and hence risk are obviously very much dependent upon cultural values and political will, which can be greatly influenced by events.[24]

Risk and its implications cannot be fully understood, appreciated, or calculated without a firm grasp of the nature, causes, and consequences of failure. While there may be a risk associated with success, in the sense that too much of anything can be a bad thing, risk is most commonly associated with failure. And since it is future failure that is at issue, the only sure way to test our hypotheses about its nature and magnitude is to look backward at failures that have occurred historically. Indeed, we predict that the probability of occurrence for a certain event, such as a hundred-year storm, is such and such a percentage, because all other things being equal, that has been the actual experience contained in the historical meteorological record. In other words, the record of occurrence of past events is a guide to determining the probability of future events of the same kind. In a similar manner, a knowledge of past failures is invaluable in estimating the probability of future failures of a similar kind in a similar context.

When there is no direct past experience of failure, as will be the case when a totally new technology is involved, risk can be based on analogy. In the 1950s, when commercial nuclear power plants were being considered as a natural extension of the nuclear propulsion technology developed by the navy, questions were raised about risk to the public. Land-based power plants would use the heat from the nuclear reaction to generate steam, which in turn would drive turbines coupled to electrical generators. Although the primary source of energy was novel, many of the generating components of the plant were very similar to those that had long been used in coal-burning power stations. Thus, conventional experience with such components as pressure vessels, versions of which would enclose the reactor, could be taken as reliable indicators of how dependable the analogous equipment would be in a nuclear plant. In other words, the record of failure of conventional components could confidently be used to predict the failure rate of nuclear-plant components. Such predictions in turn could provide an understanding of the risk—and its reciprocal, the reliability—of the projected technology on which an electric utility could base its business decision about whether or not to proceed with a project.

In like manner, the record of failures of the various mechanical and electrical components and systems that go into a power plant could provide guidance for calculating the risk of similar failures leading to a catastrophic accident that might release radioactivity into the environment. It was important for the utilities to know that risk so that they could understand their potential liability should an accident occur. When it was found that the fiscal consequences were projected to be prohibitively expensive, and no insurance company would cover all the unknown consequences of a nuclear accident, utilities balked at proceeding with commercial nuclear power development. At this point the government, which was promoting the development of civilian nuclear power, intervened through the passage of the Price-Anderson Act of 1957, which limited the financial liability of nuclear power plant operators in the event of a catastrophic accident. It was only with a well-defined cap on their own liability that the utilities were willing to go forward with the development of nuclear generating capacity. The provisions of the Price-Anderson Act have been modified and extended several times since and continue in force.[25]

Knowledge of possible ways in which a technological system can fail, and the consequences, is indispensable for making informed technical and economic decisions. This knowledge comes very much from past experience with failures, even if only by proxy and analogy, and this fact reinforces the value of case studies of failures and the accompanying lessons learned. One student of failures has gone so far as to claim that "engineers are only as good as the sum of the experience they and those who came before them have acquired." But having to master countless detailed contemporary and historical records of failures and their causes and consequences would overwhelm any engineer or designer working on advancing the state of the art. Understanding failure in more general and fundamental terms thus provides an invaluable advantage to predicting its occurrence in new technologies. But such an understanding, if it is not to be so theoretical that it is inapplicable to the real world, must arise by a form of induction from an accretion of experience-based case studies.[26]

A designer or engineer's personal and visceral understanding of failure also comes from his or her own life experience, which naturally involves the human element in an essential way. Some of this experience can be related to growing up, which of course is typically punctuated by the normal childhood accidents that lead to bloody noses and broken bones. We learn directly from these experiences the consequences of an unfortunate misstep or an ill-advised prank. It is no accident that children grow less clumsy as they age. We also learn about failure less directly but no less effectively from our more mature experiences. These often teach us by metaphor, and they can be so personal that they give us our unique take on the nature and consequences of failure. My most memorable experiences of this kind did not take place until after I had studied engineering for four years, during which time the word *failure* had evoked the concept of a grade on a test or in a course and not any understanding or misunderstanding of its nature in the context of the real world of engineered systems and their components. As I describe in the next chapter, my personal introduction to the symbolic and actual nature of failure in an engineering context came when I was a graduate student, during which time I found myself housed in a building and immersed in a culture that was centered about the symbolic and actual failure of all kinds of things.

FOUR

Mechanics of Failure

When I first encountered the concept of failure in a big way I did not fully grasp its significance, even though an overbearing symbol of it was a looming presence in my daily routine. At the time, I was studying at the University of Illinois, and during my first few years there my home away from home was Talbot Laboratory, a large red-brick building located just inside the city limits of Urbana. The lab, whose western face looks straight across Wright Street to the twin city of Champaign, dates from 1928, when it was dedicated as the Materials Testing Laboratory. Indeed, "Materials Testing" is inscribed in the stone lintel above the entrance. Ten years later, the building was renamed the Arthur Newell Talbot Laboratory, a designation then inscribed in a large stone tablet above the entrance pediment. Talbot was the professor of municipal and sanitary engineering who in 1890 had been put in charge of theoretical and applied mechanics at Illinois, where the discipline and later the department that housed it in Talbot Lab came to be known affectionately by the acronym TAM.[1]

Materials testing is the alpha and omega of structural engineering analysis and synthesis. When engineers design a bridge or a building, they have to know the strength of the steel or concrete that is specified for the beams and columns that hold it up. The ultimate strength of a material is effectively its breaking point, and if the material used is not a standard one whose properties are tabulated in a handbook, they can only be determined with confidence by testing samples, which effectively means pulling them apart or crushing them to the point of failure. Once this is done and the data are recorded, the theories, equations, and calculations of theoretical and applied mechanics enable engineers to predict, with considerable accuracy, how much stress a real structure can withstand and thus how much load it can bear. This proactive failure analysis known as design is what turns tentative plans into final ones. Similarly, when a structure collapses unexpectedly, a postmortem failure analysis to determine its cause will rely on the analytical tools of theoretical and applied mechanics to check if the structural design was done properly. Those tools, in conjunction with the proper testing equipment, will also subject to failure samples of the material from the bent, twisted, and torn parts scattered throughout the wreckage to establish if they indeed had the strength they were presumed to have had.

According to the philosopher of science Ernst Mach, mechanics is "at once the oldest and the simplest" branch of physics. Traditionally, mechanics has concerned itself with forces and how they affect the equilibrium or motion of masses ranging from atomic particles to solar systems. To this day, it remains an anomaly that at Illinois the physics department, where engineering students first encounter university courses incorporating mechanics, is administratively a part of the College of Engineering. And the Physics Building, like Talbot Lab, is on the engineering part of campus, which means it is north of Green Street. This does not indicate that the university's proud physicists have any less of a sense of universality of purpose or of superiority over their engineer colleagues than physicists elsewhere. Nevertheless, when asked to vote on whether they would like their academic home to be relocated to the College of Liberal Arts and Sciences, the physicists chose to remain with the engineers. They thus continued a curiously balanced asymmetry at Illinois, where the chemical and biomolecular engineering program is in with the arts and sciences. But the administrative division that anomalous academic departments and programs call home is less important than what they do there. At Illinois, most of the physicists do what they do in physics departments just about everywhere.[2]

In modern physics (and astronomy), there are so many sub-branches of mechanics that they have to be qualified to be distinguished. Roughly but not exclusively in order of the scale of their subject matter, these include: quantum mechanics, statistical mechanics, classical mechanics, relativistic mechanics, and celestial mechanics. But theoretical and applied mechanics, whether at Illinois or elsewhere, is considered a part not of physics but of engineering. The objects of its study are those found not in the given universe but in the world of designed and made things—the world of engineering. Thus faculty members and students in the TAM department that I knew developed and studied theories about structural materials and the objects made of them, like beams and columns and other generally large components. They also studied the environments—like water and dirt—with which engineered things interacted, and how the theories were applied to engineering problems, like the behavior of buildings and bridges. The names of some of the sub-branches of engineering mechanics suggest its interests in down-to-earth things. In no particular order, some of the sub-branches are: structural mechanics, soil mechanics, fluid mechanics, solid mechanics, and biomechanics. Though mechanics in physics and in engineering have common roots, their practitioners have divergent interests and objectives. (Needless to say, auto mechanics is a branch of neither physical nor engineering mechanics.)

In the 1960s, when I roamed them, the corridors of Talbot Lab were hung with portraits of its namesake and other prominent and historical figures in mechanics and hydraulics, which is the mechanics of flowing water. But the portraits were not what caught my attention when I first set foot in what was then to me an unconventional building. Its "halls" were unusual in that most of them opened into offices and classrooms on one side only. On the first floor, the side of the hall opposite the office and classroom doors had no wall and overlooked a great machinery room that reached from the open basement floor to the exposed roof trusses. On the second and third floors, the side of the hall bordering the great open room was defined by a doorless wall punctuated by windows that overlooked the large interior space, as if they were put there to enable sidewalk superintendents to keep track of what was going on down in the great interior destruction site. The building had these half-halls on three sides of what might be described as a great central atrium, with the fourth side being closed in by an outside wall containing many large south-facing windows. This arrangement came about long before the modern architectural feature of a soaring hotel atrium became fashionable, and Talbot Lab's enormous open space contained not fountains and foliage but heavy mechanical equipment for testing full-size steel and concrete structural components. These parts naturally had to be moved in and out of the machines on the floor below, and this was done with the aid of a large overhead crane. The laboratory space was in fact a massive crane bay, which is exactly how we all referred to it.

Shortly after Talbot Lab was completed, a three-million-pound-capacity Southwark-Emery Universal Testing Machine, made through a partnership of the Baldwin-Southwark Corporation and the A. H. Emery Company, was installed on the basement floor of the bay. As with just about every major industrial enterprise, there was a story behind the founding of the testing machine company. The short version is that the Baldwin Locomotive Works, which had been established in Philadelphia in 1831, even-

A massive testing machine, capable of exerting three million pounds of force to break large structural parts made of concrete and steel, was long the visual centerpiece in Talbot Laboratory at the University of Illinois. The behemoth, which towered over the space it commanded and dwarfed workers tending it, was a symbolic reminder to engineers of the ever-present possibility of failure occurring in their designs.

tually designed and developed machines to test to failure some of the large parts that went into its steam locomotives. In time, Baldwin-Southwark, working with the materials testing innovator Emery, began to make testing machines for sale outside the locomotive works.[3]

The Southwark-Emery machine installed in Talbot Lab dominated its interior space the way the Corinthian columns do that of the Pension Building in Washington, D.C., which now houses the National Building Museum. In Urbana, the enormous machine towered over everything, as if it were the signature display at a great industrial exposition, rising up four stories and from certain angles appearing to mingle with the exposed trusswork that supported the lab's slightly peaked roof. The giant testing machine mostly just sat there as a static structure, a tall and husky stabile whose parts were painted the almost-Calder-like colors of bright yellow, bright red, and battleship gray, the colors designating, respectively, the giant jaws through which force was applied, the work platform-elevator that carried the technicians up and down, and the structural frame that supported everything. When there was any activity around the machine, the technicians appeared to be tending to its needs rather than adjusting it for its ultimate purpose, which was to break things too large to be broken by other laboratory means. Engineers naturally have to understand the strength of the specialized structural components out of which they design and build things. Often they can do so theoretically, based on the strength of material samples, but when something is too complex in form or too innovative in fabric, the only way to be sure of its strength is to test it physically to failure. Even today, when computer models have become ubiquitous, large-scale testing is necessary to validate and calibrate complex models of complex parts.[4]

The dominant feature of Talbot Lab and, by extension, of the TAM department itself was thus a massive reminder of the central role that failure played in the discipline. The great testing machine stood, symbolically and actually, at the ready to apply real forces to confirm or contradict theoretical predictions of strength. It was the final arbiter between competing theories and contradictory predictions. The biographical notes appended to the portraits of notable scientists and engineers—mechanicists all—overlooking the crane bay stressed their successful achievements, but the lurking stabile that could be transformed into a quasi-static mobile was always there to render a verdict whenever notables and would-be notables disagreed. Failure was the incontrovertible counterexample to putative success, and it was at the implicit heart of everything for which TAM stood. This would not become evident to me until years after I had left Urbana with my sheepskin, wearing it like a wolf looking to move undetected among a flock. Theoreticians such as I was then dreamt of pouncing on what we considered sheepish experimentalists with an equation that made all their tests unnecessary; we interlopers were tolerated by the experimenters, who knew that their own browsing among the materials of the field would yield failures that were not predictable by us or by anyone. As one student of science and technology has observed, "scientific understanding didn't progress by looking for truth; it did so by looking for mistakes." Analogously, engineering understanding did not progress by looking at successes; it did so by looking at failures. This same idea with a new twist was echoed by a corporate executive speaking on the subject of technological innovation: "Understanding comes from failure; success comes from understanding failure and acting upon this knowledge." This echoes the old Chinese saying that "failure is the mother of success."[5]

Scale-model tests to failure are frequently employed in engineering, but sometimes only a full-scale test will do. It was hard to miss when something big was being readied for testing in the crane bay. But it was not always easy to tell just by looking whether the lower set of jaws of the gigantic thing, which moved excruciatingly slowly, was tearing a specimen apart or just at the ready to do so. To alert the occupants of Talbot Lab that something was about to fracture—releasing enormous amounts of energy in the form of an explosive noise accompanied by a recoil of the megamachine that made the entire building shudder—an alarm was sounded whenever a test was headed toward the breaking point. The whole of TAM pivoted on the test to failure, but most of the theoretical graduate students shut their ears to the noise and most of the applied students up on the third floor, where smaller equipment was housed, looked down their noses at the large machine.

The monster was sometimes used to smash things much smaller than itself, though not everything it destroyed was small. Those objects condemned to the fate of being crushed like an empty aluminum can underfoot were placed between the machine's moveable platen and the hard basement floor. Among the most common uses of the monster in this manner was to test the strength of large concrete cylinders. The cylindrical specimen is a standard in concrete testing, for it is relatively easy to pour the wet material into a can-shaped mold and let it cure for an appropriate time. (It also is suggestive of a structural column, one of the principal applications of the material.) By crushing the cylinder in a testing machine until it fails, engineers can determine the strength of the cylinder and hence of the material of which it is composed. For modern concrete mixes this ultimate strength can range from about 2,500 to greater than 10,000 pounds per square inch, depending on the ingredients, their proportions, and the age of the specimen. For common construction projects, like an apartment building or a parking garage, a standard-size test cylinder is six inches in diameter by twelve inches high, a little larger (and, at about thirty pounds, a lot heavier) than a can of tomato juice. The force needed to break the cylinder can range from about 70,000 to almost 300,000 pounds. This can be done with a machine compact enough to fit into a small laboratory room. The cylinders are formed from the same batch of concrete that is placed in the structure under construction. They are stored under controlled conditions and tested at intervals to monitor the strength of the concrete as it ages in the actual structure.

For a major project, such as a large hydroelectric dam, a concrete test cylinder has to be much greater than six inches in diameter because of the size of the aggregate used in the concrete mix. Aggregate, typically in the form of stone or gravel, is added to the mixture of cement, sand, and water to bulk it out and thereby reduce the required amount of the most expensive ingredient—the cement. Materials used for aggregate can range from rounded river rocks and stones to angular pit gravel, with their sizes depending upon the scale of the structure. In ordinary construction, maximum aggregate size might be about a half inch, but in a large dam it might be greater than six or eight inches, or larger than the standard-size test cylinder.

A rule of thumb is that the diameter of the test cylinder should be at least three times the diameter of the maximum aggregate size. In Hoover Dam, this was about eight or nine inches, and so test cylinders had to be about two feet in diameter, more the size of a fifty-five-gallon oil drum than a tomato juice can. Just moving the concrete cylinders around took work, for they could weigh more than three tons each. It would require in excess of a million pounds of force to break a cylinder of that size, if it was made of ordinary concrete, and that is what big machines like the one in Talbot Lab sometimes were called upon to do. If the cylinder were made of high-strength concrete, say one capable of bearing on the order of 10,000 pounds per square inch, then the force to break it might exceed four million pounds—clearly beyond the capacity of even the Talbot Lab machine. Some archival cylinders made with the concrete used in Hoover Dam were tested when they were about sixty years old, and the material was found to have a strength in excess of 9,000 pounds per square inch, which means that a force well in excess of three million pounds would have been needed to test them to failure. That is why four- and five-million-pound testing machines were located in the Bureau of Reclamation laboratories, where the necessary tests could be conducted. Those of us graduate students who did not work directly with concrete were blissfully ignorant of such facts. To us, the testing machine in the crane bay was the ultimate symbol of strength.[6]

Nevertheless, we seldom talked about the three-million-pound gorilla in the crane bay, except to refresh our memories about its impressive specifications in preparation for showing a visiting relative or friend around Talbot Lab. To them, our pages of equations were nothing compared with the tower in the bay. We veteran graduate students may have been able to ignore the gorilla, but guests seeing it for the first time could not. They wanted to know its name, and how much it weighed, and how powerful it was. Yet all we could tell them was that it was a Southwark-Emery and could pull with the force of ten thousand horses, or at least did when it was young. Rumor had it that during one particularly strenuous test it had broken an anchor bolt and so no longer had the capacity that it once did. Now, shackled to the basement floor and caged in by mechanical halls of fame, the monster was a spectacle upon which everyone looked, but few saw its true significance. This embodiment of past and future failure was in fact the raison d'être for the crane bay, for Talbot Laboratory, for the Department of Theoretical and Applied Mechanics, for its faculty, for its students, and for their research and development. Without the ability of the great machine and its smaller counterparts to break the toughest nuts that could be placed in or beneath their jaws, those of us who would eventually study how and why things that once were successful structures did in time fail would have no confirmation of so many of our theories and applications. But perhaps because it mostly just sat there so sullenly, the gorilla in the crane bay was something that we seldom talked about back then.

My first office in Talbot was in the northeast corner of the third floor, which was about as far from the centerpiece testing machine as it was possible to get and still be inside the building, but I could nevertheless hear the report of riveted steel pulled asunder and feel the bounce and subsequent vibrations that it set up in the building. I shared the subdivided office with three other graduate students, who worked as teaching assistants, and a new assistant professor, who for lack of space elsewhere was exiled to our remote corner of the department. Most of us students were at our desks from morning till night studying, preparing lectures, or grading exams. Even if they did not share offices, many of the graduate students and some of the faculty members housed in Talbot fraternized regularly, foraging across Wright Street for coffee, lunch, and late-night beers. It was principally on those semisocial occasions that those of us in theoretical mechanics learned a little bit about applied mechanics.

Among the faculty members who joined us most frequently at the Capitol Bar and Grill, located around the corner on Green Street, was JoDean Morrow, variously known to graduate students as JoDean, JoMo, or simply Jo. Like many of the TAM faculty, JoDean was a Midwesterner by birth and by education. He was a native of Iowa, but at about ten years of age his family moved to Indiana. He studied civil engineering at what was then known as Rose Polytechnic Institute, located in Terre Haute. He received his B.S. from Rose Poly in 1950, worked for about a year with the Indiana State Highway Commission, served in the army from 1951 to 1953, and then enrolled as a graduate student in TAM. For his Master's thesis research, he used the giant testing machine in the crane bay to collect data on full-scale reinforced-concrete structural frames; he employed smaller testing machines for his doctoral research on cyclic loading of small specimens. After receiving his Ph.D. in 1957, he became an assistant professor in TAM. When I was a graduate student, from 1963 to 1968, at least half of the faculty seemed to have done their graduate work in TAM and stayed on to work and teach in Talbot Lab.[7]

Out of the lab, JoDean often seemed to be trying to shock or even to scare young graduate students. After one late-night session at the Capitol, he offered to give me a ride home in his sports car —an MGA, I later learned—something I had not previously experienced. Not surprisingly, the car was loud and it cornered left and right the way I imagined a true sports car would, resulting in centrifugal forces that pushed and pulled my body right and left against the door and toward the driver. JoDean kept going faster and faster around corners and kept asking me if I had had enough. There were no seatbelts, naturally, and I did worry that the door I was pressed against would fly open. But the last thing I wanted was to appear to be a wimp from New York City, where it was possible to grow to maturity without owning, driving, or even riding in a car, let alone a sports car. So I was determined not to let him see my breaking point; I told him that I was doing fine. At that, he drove into a large empty parking lot and began to peel rubber, fishtail, and skid every which way, but I held on and just hoped we would not flip over or attract the attention of the police. I guess I passed the test, because after that ride, Jo seemed to be warmer toward me; I also had an increased fascination with him and his lifestyle.

One night, instead of asking me if I wanted to take a car ride, JoDean invited me up to his office and lab to show me the kind of work being done there. The Materials Engineering Research Laboratory occupied space in the northwest corner of the third floor of Talbot, a short walk physically from my side of the building—albeit one that passed by the crane bay—but a long journey intellectually. My office was in what I considered the theoretical wing of the department. When I was not taking courses in continuum mechanics and shell theory in Talbot Lab, I was taking courses in partial differential equations and complex variables in Altgeld Hall, the home of the mathematics department, located just across Green Street. I and my like-minded colleagues took no courses in, and at the time had no interest in, experimental techniques or materials engineering.

As theoreticians usually do, we felt superior to the experimentalists, who we thought merely applied our theories. In fact, it would often be the other way around, with the experimental work informing the theories. JoDean, who liked to expound on human psychology and social behavior, certainly understood this and seemed to want to dissuade us of our prejudices. He tried to remind us that "theoretical mechanics seeks exact answers to approximate problems, while applied mechanics seeks approximate answers to exact problems." Although it would take a while for me to realize the full implications of that truth, in the meantime I developed a guarded respect for what JoDean did in the laboratory. What he showed me in his lab—located in the space once occupied by the historic fatigue-of-metals laboratory that had mechanically rotated steel beams for as many as a billion cycles to simulate what happened to railroad axles—were state-of-the-art machines that were computer-controlled and could apply complicated loading patterns to test specimens and automatically count the number of cycles it took for them to fail in a process known as low-cycle fatigue. The table-top machines were miniature compared with what was standing in the crane bay, and I was impressed not only by the sophistication of the hardware but also by the thoughtfulness of the experimental protocol that JoDean explained to me. He clearly had a profound knowledge of the behavior of materials and how specimens made of them failed, yet he was humble in his admission that there was so much more yet to be learned. Hence the experimental program that he and his stu-

Instruments of failure housed on the third floor of Talbot Laboratory were in
stark contrast in scale and agility to the monster in the crane bay. In this photo,
taken circa 1974, Professor JoDean Morrow (1929–2008) is showing off a new
table-top computer-controlled testing machine. Looking on are (left to right):
Richard Shield, head of the theoretical and applied mechanics department at
Illinois; Herb Johnson, representing MTS Systems Corporation, the manufacturer of the machine; and Daniel Drucker, dean of the College of Engineering.
Seated at the far end of the table, beside the computer that controlled the
machine, is Kelly Donaldson, a graduate student.

dents pursued, and that visitors from around the world came to
Urbana to learn about.[8]

Graduate-student offices and officemates were frequently reas-
signed within TAM, which we pronounced "tam," as in "tam-o'-
shanter," the woolen cap worn by Scotsmen. Some of the senior
professors insisted on using the abbreviation T&AM, which was
unpronounceable as a single word and so was not a true acronym. In any case, as old research groups were dissolved and new
ones formed, students—but generally not faculty members, who
seemed to have squatter's rights to their offices, until they could

lay claim to a larger one—were moved around the building. My second office was in the southeast corner of the third floor, with a group loosely allied with a program in mechanical vibrations, those shaky motions that accompany earthquakes and can move structures to failure. I remember mostly discussions of politics in and around that office, especially with another graduate student widely known to be a staunch and vociferous supporter of Barry Goldwater, who was running for president at the time. My stay in that office was brief, which suited me fine, since my desk was in the middle of a large, open central space and I had found it difficult to find quiet time to study there.

My next office was more to my liking. It was in the southeast corner of the first floor of Talbot, in an area occupied by the concrete lab. To get to my new office, I had to pass through an anteroom that normally housed a receptionist and secretary, but their desks were empty because the research program that had paid their salaries was between funding grants. Faculty offices opened off this anteroom, and a large partitioned office for graduate students was in a back corner. The choicest compartment was designed for a single occupant, and it had long been claimed by the senior graduate student in the group. I was assigned a desk in a large three-person compartment, which I shared with two other graduate students who were also teaching assistants and so had no natural research group with which to be housed. One was an Italian who wrote letters to the editors of newspapers and news magazines and was especially proud of one that was once published in *Time*. He wrote his dissertation on thin cylindrical shells—aluminum beverage cans and rocket casings were relatively new at the time, and there was still much research to be done on their behavior—and went on to work in Vienna for the International Atomic Energy Agency. My other office-compartment mate was a West Virginian who was married with kids and so usually stayed home in the evenings. He wrote his dissertation on the stability of a deformed elastic shell—such as the domed bottom of an aluminum can—and went on to teach at the University of Maryland. We were a diverse but compatible group because we generally kept distinctly different hours, and when we were all in the office at the same time we largely studied back-to-back-to-back in silence.[9]

As is the tradition in many a research lab, in our enclave there were regular coffee klatches at which graduate students and faculty mingled and talked about their work and ideas. The concrete group's coffee breaks took place in a laboratory room full of chemistry benches and equipment, and it was there, amid test tubes and centrifuges, that I frequently engaged in debate with my colleagues over distinctions between theoretical and applied mechanics. I had been introduced to the history and philosophy of science as an undergraduate and had continued to read in those areas in my spare time as a graduate student. These interests meshed nicely with the axiomatic approach to continuum mechanics that would attract me to my advisor, Don Carlson. He, a Midwesterner who was born in Tampico—the small village in northwestern Illinois that was also the birthplace of Ronald Reagan—had been an undergraduate at Illinois, and received his B.S. in engineering mechanics, as the undergraduate TAM major was called. After his own stint in graduate school, he had recently returned to the department as a very young assistant professor with a fresh Ph.D. from Brown University, where engineering and applied mathematics were very closely linked.

Continuum mechanics as Don had learned and taught it was virtually indistinguishable from mathematics. As its name implies, continuum mechanics deals with the interrelationships between the forces, motions, and changes of shape of bodies considered to be made up of contiguous parts, as opposed to the physical systems comprising such discrete parts as atoms or billiard balls or planets. All the time I was in graduate school, it was the height of fashion in the field of continuum mechanics to rationalize its

Donald E. Carlson (1938–2010), for forty-two years a professor of theoretical and applied mechanics at the University of Illinois, was the author's dissertation advisor. Don's interests and talents were on the extreme theoretical end of the spectrum that TAM spanned at Illinois, and the word *failure* was virtually absent from his professional vocabulary.

foundations, which meant infusing it with a rigorous mathematical structure. The way Don expounded on the subject was to write on the blackboard axioms, theorems, lemmas, corollaries, and the like, and then to work out proofs of them as if we were taking a course in mathematics, concluding with the mathematician's assertion of QED, the Latin declaration *quod erat demonstrandum,* "that which was to be shown." The proofs were carried out with strict mathematical logic, and what we thought to be the best of teachers and students followed them to the Greek letter. Missing from the approach, however, was usually an appeal to experimental evidence as the basis and motivation for the mathematical manipulation.

I do not recall the word *failure* being used much in lectures, papers, or books on continuum mechanics. Mathematical theorems did not fail, for they were written up on the blackboard and copied down in student notebooks in a daze of predetermined correctness, completeness, and conclusiveness that was infectious. How beautifully everything followed mathematically from hypothesis through proof to conclusion. The lecture, the lesson, the notes were hermetically sealed off from the laboratory, the test, and the shock of structural steel parts breaking in the crane bay. The closest we got to the concept of failure was to disprove a hypothesis by counterexample. The words used in such exercises were associated with mathematical symbols that were identical to those used in physics and engineering—force, mass, acceleration, energy—but for me at least they were more names for the symbols rather than designators of the concepts.

One conundrum that baffled those of us engaged in continuum mechanics was what to call ourselves. If we said we were mechanics, then we risked being mistaken for manual workers, artisans, machinists, repairers of machines, workers in the crane bay. The word *mechanician* was favored by some, but it seemed to be just a fancier and more Latinate form of mechanic. Some history-minded pedants suggested that we were *geometers,* but that seemed to connote a focus on space to the exclusion of time. Besides, the word was virtually obsolete and called to mind more the field of geography than mechanics. We could call ourselves engineers, but continuum mechanicians—to use a cumbersome but unambiguous option—seemed constantly trying to distance themselves from engineering, which had its own awkward connotations, with the word *engineer* evoking an image of a railroad-train driver wearing a blue-and-white striped cap that looked like it was made from pillow fabric.

Some theoretical engineers—the amusingly ambiguous term suggesting that those who engaged in theoretical musings were engineers in theory only—went so far as to call themselves scientists. Indeed, the academic system aided and abetted this approach by designating many an engineering degree as Bachelor of Science. Other theoretical engineers identified themselves as research engineers, thereby distancing themselves from locomotive drivers as well as from design engineers. They were clearly affiliating themselves with the leading end of research and development, the trailing end perhaps being perceived by them as too close to design. Research was generally associated with successful endeavors—those that were not successful were just not mentioned. Development must confront the moment of truth, where the real-world implications of a new theorem may or may not be achievable. Scientists succeed; engineers fail. But we were ignorant of Lord Kelvin's observation that "the steam engine has done much more for science than science has done for the steam engine."[10]

All the time I was a graduate student, I relied upon teaching assistantships to support myself. This meant not only that I owed my allegiance to no particular research group but also that I had no particular research-based home office. Wherever I was housed, it seemed, I was in foreign territory on a visa. Still, when my office was in the transitioning concrete lab, I naturally gravitated toward my neighbors' coffee breaks and conversation. Among the topics debated amid the laboratory benches, chemical apparatus, gravel sieves, tabletop testing machines, and ubiquitous cement dust were the nature of force and the status of Newton's Laws. Mostly, we talked past each other. We failed to communicate. When we talked about failure, it was in the personal context of failing a course exam, a qualifying exam, a preliminary exam, or a final exam. In fact, failing any of those exams was a far rarer event than talking about failing them.

But being housed in the concrete lab did expose me to another aspect of the applied side of TAM and also to engineering as opposed to mechanics. Since concrete research was of great interest to civil engineers, many of the students and faculty affiliated with the program had a foot in each discipline. The civil engineering department at Illinois was widely known and highly regarded, largely because of its many distinguished faculty and successful alumni. The distinctions between mechanics—whether theoretical or applied—and engineering were sometimes subtle, especially in the classroom, but they were dramatic outside the academy. Thus, while TAM and civil engineering students could use the same words when engaged in technical conversation, their ultimate interests in the subject matter were worlds apart. The mechanics people seldom called themselves engineers, and often seemed to think of themselves as scientists, or at least as engineering scientists. The engineers, by contrast, were proud of and comfortable with the way they were labeled.

With the growth of TAM, largely manifested in the addition of new faculty members, space in Talbot Lab became extremely tight, and graduate students like me, with no pressing need to be near experimental equipment, were exiled across the street to a building known, perhaps appropriately, as the Woodshop. In fact, its name derived from an earlier practical function, but the postwar engineering curriculum had evolved into a more science-based one, leaving little need to devote a whole building to woodworking equipment for furniture-, form-, and model-building. The front part of the Woodshop had been converted into classrooms and graduate-student offices. It was there that I would spend the remainder of my days at Illinois, venturing into Talbot Lab only to check my mailbox, attend seminars, confer with my advisor, and meet up with someone with whom to walk across Wright Street for coffee or beer.

For all of my informal exposure to fatigue and concrete research at Illinois, I remained firmly entrenched in the theoretical camp for almost another decade, publishing the results of my mathematical manipulations in what I considered distinguished journals and building what I thought to be a respectable curriculum vitae. But there was a constant nagging in the back of my mind that as much as I thought I understood engineering science I did not fully understand engineering, the subject that I was by then being paid to teach at the University of Texas at Austin. What pulled me away from my strict diet of equations was what I learned at Argonne

National Laboratory. When I was invited there to give a seminar on my work, I found myself doing so before researchers in the Materials Science Division. It was soon clear that they could follow my mathematics but I could not fully respond to their questions rooted in a deep understanding of the material stuff of the real world—and how it failed. They wanted to know how the mathematical quantities that I was talking about and manipulating with apparent ease related to real materials that were not nearly so easily manipulated as the factors of an equation. Without laboratory or real-world experience, I did not have a satisfactory answer, at least not one that satisfied me.

At the time, one of Argonne's principal missions was to develop a liquid metal–cooled nuclear reactor that would breed more fuel than it consumed in the process of operating. This was clearly a real engineering problem, whose solution required the cooperation of nuclear physicists, materials scientists, fluid mechanicians, structural engineers, and a host of other specialists housed in various buildings located around the inner ring road on the laboratory site. Unlike in the university environment, where individual groups of faculty members and graduate students worked on separate problems and seldom interacted, at a mission-oriented laboratory there was a natural and pressing need for true interdisciplinary interaction. Success and failure were group experiences.

The idea of working on a common problem with a wide variety of scientists and engineers appealed to me, not least because it would give me the opportunity to see how science and engineering related to each other. A nuclear reactor research, development, and demonstration program seemed to be a natural venue for bringing together the two sides of theoretical and applied mechanics. When I was offered a position at the lab, I quickly accepted it; it presented me with opportunities for learning that at that stage in my career the university had not.

My official title at Argonne was mechanical engineer, and I was based in the Reactor Analysis and Safety Division. As I understood it, the division was responsible for analyzing the nuclear activity taking place in the core of a reactor and understanding the implications for safety of various anomalies—failures, really—that might take place in an operating plant. (Like NASA, the nuclear industry abounded in euphemisms for "failure.") Those anomalies could involve anything imaginable, from the inadvertent removal of a control rod to the sudden rise in temperature that would accompany an interruption of coolant flow. This last scenario was likely to happen if, for example, a small imperfection in one of the main coolant pipes grew into a crack that weakened the pipe to such an extent that it gave way to a spontaneous break of sizeable proportions, leaking coolant faster than it could be supplied—a clearly dangerous anomaly, or failure mode. There was also much attention paid to the so-called hypothetical core disruptive accident, known as an HCDA, which was the most-feared failure scenario. It was imagined to be possible if a sequence of failures led to molten nuclear fuel interacting with liquid sodium to produce a damaging gas bubble—in other words, an *explosion* of sorts, though that word was not used. As a concept, an HCDA straddled the theoretical and the real by being at the same time a hypothetical construct and an accident with potentially real consequences. One of the key questions was whether the steel pressure vessel in which the reactor core resided would break open during such an event. Everywhere I looked at Argonne, I saw a focus on failure and the problems it posed.

Since no one on the division's staff specialized in the analysis of cracked and broken pipes, I was charged with defining and leading a new group in the area of fracture mechanics. In 1975, when I moved to Argonne, this was a relatively new field, involving a subject that was not yet widely taught in universities. There had been concerted research into the subject since mid-century, and theoretical work was published in the *Journal of the Mechanics and*

Physics of Solids, which started up in 1952. But *Engineering Fracture Mechanics,* a more applied journal, dated only from 1969, and the *International Journal of Fracture* from 1973. In the mid-1970s, the first monographs and textbooks on the subject were just beginning to appear. To me, these were all indispensable resources for learning about a topic that had not even been mentioned in graduate school.[11]

A characteristic of early monographs and textbooks on fracture was the seemingly obligatory introductory section or chapter on historical background. This impressed me greatly, for it showed that failure had long been the bane of engineering; moreover, it emphasized to me that the concept of fracture did not derive from theory but rather the other way around. Thus the opening chapter of Rolfe and Barsom's *Fracture and Fatigue Control in Structures,* first published in 1977, began with a historical overview of some landmark failures due to brittle fracture, a phenomenon in which a structure or machine part gives virtually no warning that it is about to fail—it just suddenly snaps apart or breaks open. The cases included the 1919 Boston molasses tank failure, in which twelve people died and forty were injured when more than two million gallons of the sticky syrup were released spontaneously; the over two hundred welded Liberty ships built during World War II that suffered serious brittle fractures, some resulting in a ship's suddenly breaking in two; the de Havilland Comet early jet airliners that mysteriously broke up in flight catastrophically in the mid-1950s, leaving no survivors; and the Point Pleasant Bridge, which collapsed suddenly in 1967, throwing commuters into the Ohio River and killing forty-six of them. Such dramatic and tragic failures made it clear that fatigue and fracture were not merely academic pursuits.[12]

That is not to say that theoretical and applied research was not important in advancing our understanding of what made cracks grow and structural failure occur. Indeed, the brittle fracture of steel ships inspired research into the chemical makeup of structural materials and into the nature and effects of welding, especially as it might affect fracture resistance. To me, coming to the field in the mid-1970s, the most prominent people dealing with fracture were affiliated with the American Society for Testing and Materials (ASTM), an organization founded in the late nineteenth century to deal with the chronic problem of railroad rails and axles breaking. Shades of Talbot Lab! My introduction to ASTM came through Rolfe and Barsom's frequent references to the society's Special Technical Publications, volumes that collected peer-reviewed papers on a specific topic presented at a specialist meeting. Today known as ASTM International, the society bills itself as "one of the largest voluntary standards development organizations in the world." Rows of books of annual ASTM Standards covering diverse products and materials—ranging from pencil lead to structural steel—once a familiar sight on the reference shelves of technical libraries, have largely been replaced by digital versions, but the work of the thirty thousand ASTM members proceeds largely through face-to-face committee meetings, including that of Committee E08 on Fatigue and Fracture, which obviously concerns itself with failure.[13]

After the 1979 accident at the Three Mile Island Nuclear Generating Station in Pennsylvania, the bottom fell out of nuclear reactor research and development at Argonne, and there was a mass exodus. When I relocated to Duke University in 1980, I was at a crossroads in my career: I was hired to fill a vacancy in the area of theoretical continuum mechanics, but my research interests had become rooted in applied fracture mechanics. I inherited an underutilized structural engineering laboratory, which had been designed and built about 1960 to test full-scale reinforced-concrete beams to failure. The lab had a "strong floor," which meant that large testing equipment could be anchored to it and could exert very large forces on whatever—also anchored to the floor—was being tested. (The floor effectively served as a giant testing machine lying on its side.) In the two decades since the lab had been constructed, large-scale testing had fallen out of favor at Duke—as it also eventually would at Illinois itself, in spite of its distinguished tradition there—and the strong floor supported smaller testing machines that were computer-controlled and capable of operating at high rates of loading for impact testing and at high frequencies for fatigue experiments. Some of these machines, which were descendants of the equipment Jo Morrow had introduced me to a decade and a half earlier, were being used by faculty and graduate students in biomedical engineering who were interested in the mechanical properties of human tissue. My graduate students used similar machines for experiments on the growth of cracks in and the fracture of concrete.

In the meantime, I had developed an increasing interest in failure in more general terms. Highly publicized failures, both historical and contemporary, understandably have long captured the attention of engineers and non-engineers alike. In addition to the fracture- and fatigue-related failures, there were those rooted in design. The Tacoma Narrows Bridge, whose twisting and failing deck was captured on film in 1940, and the 1981 collapse of the elevated walkways in the Kansas City Hyatt Regency Hotel were failures crying out for explanations involving human as well as technical factors. Explaining diverse failures such as these called for a new kind of approach, one that was not generally employed in mechanical testing laboratories.

My attempt to explain such failures and why they occur in the first place resulted in the book *To Engineer Is Human,* in which I employed a wide variety of case studies, ranging from broken toys to fallen bridges, to illustrate general principles. Failure of all kinds continued to interest me, from the breaking of pencil points to the concept that form does not follow function so much as it does failure. Such themes played a significant role in motivating my books

The Pencil and *The Evolution of Useful Things*. Some years later, on the basis of such books, I was pleased to be invited back to Urbana to receive a distinguished alumnus award. On the occasion of that visit, I was asked to give a talk in Talbot Lab. I welcomed the opportunity to lecture again in the building where had I taught my first class—in a course offered exclusively to architectural engineering students—and I chose to speak not on the mathematical theories that I had so immersed myself in while a graduate student but on the paper clip as a case study in the failure-driven evolution of artifacts generally. With that lecture, I declared to my former teachers that my transition from theoretical to applied mechanics had been complete.[14]

It was a bittersweet moment. The voluminous crane bay had been filled in with what was in effect a building within a building, in order to provide more offices, some bench-top laboratories, and new lecture rooms. The once-expansive three-million-pound testing machine was now visible from the hallway only through a small window in the new wall that had been built to enclose the former atrium space. Talbot Lab had finally gotten conventional two-walled hallways. The room in which I lectured was handsome and appeared to be comfortable, but it was somewhat disorienting for me to reflect on the fact that I was standing in a location that used to be open space through which air horns were sounded to warn the denizens of Talbot that soon a very large structural part made of steel or concrete was to fracture and the ground and building would shake and shudder in the wake of the induced failure. Now, about two decades later, the Department of Theoretical and Applied Mechanics itself is no more. In 2006, TAM was merged with mechanical engineering at Illinois to form the Department of Mechanical Science and Engineering. I wonder if the graduate students in that new department discuss the order of and the difference between the two components of its name. And I wonder how they acquire their personal sense of failure.

FIVE

A Repeating Problem

Recently, while visiting a new dental office to have my teeth checked and cleaned, I was reminded of how much overlap exists in the technical interests of dentists and engineers, especially structural engineers. When the dentist asked me my occupation, I told him that I taught engineering. His response was that dentistry was a kind of oral engineering, to which I nodded and grunted agreement. He also offered that his son almost studied engineering in college, but in the end decided he wanted to be a dentist. This non sequitur reminded me that one of my former students, who wrote a term paper on the famous Cooper River bridges of Charleston, South Carolina, and eventually coauthored a book on the original structure, went to dental school after practicing engineering for several years. I do not know any dentists who left their practice to study engineering.[1]

Regardless of their professional roots, dentists and engineers alike have to know a good deal about mechanics, materials, and maintenance. They have to understand the forces at play on a bridge, whether it be between two sound teeth or between two soft river banks. Dentists and engineers have to understand the holding strength of amalgams and epoxies and of steel and concrete. And they have to understand the importance of limiting deterioration, whether that means practicing prophylaxis to control dental caries or painting steel to prevent corrosion. But the growth of decay and corrosion is insidious. Thus it is perhaps in promoting regular checkups that dentistry has the most to offer engineering, in the sense that it makes evident even to the layperson how important regular and thorough examinations and inspections are to the health of our oral infrastructure and, by analogy, our national infrastructure. In a routine dental checkup, the structures in our mouth are gone over with picks and probes to detect the small irregularities that can be harbingers of worse to come, to find the imperfections that can lead to cracks and the cracks that can lead to spontaneous fractures, to locate and fill cavities that, left unchecked, can threaten an entire tooth. Every year or so, X-rays are taken, and suspicious areas compared with benchmark images from an earlier visit. Given that we all have—or at least once had—teeth, we know the importance of regular dental examinations and prophylactic care.

Indeed, we all know, perhaps by having learned the hard way, that neglecting our teeth can lead to their deteriorating beyond repair. We seem not yet to have fully learned that the same applies to our bridges, roads, and other public works. But going to the dentist can get us thinking about a lot more than our teeth. The experience can remind us of knowledge and lessons about good engineering practice that are forgotten at our peril. When we go to the dentist with a broken tooth, we are likely to learn that the tooth probably had had a crack in it. The crack may have gotten started quite a while back, maybe when a sharp blow from a mating tooth nicked the enamel but so slightly that even a dentist could not easily detect it. Over time, as we chomped on three meals a day, plus snacks, a little crack grew ever so slowly into a larger one. In mechanical terms, the forces imparted to the tooth in the course of chewing alternately opened and closed the crack and in the process extended it ever so slowly further into the tooth. As long as we did not chomp our teeth together too suddenly or use them as if they were a pair of pliers, the crack was stable and the tooth was safe. Eventually, however, the crack would reach a point where little force was required to get it to fast forward like a crack in a window pane tapped ever so gently with maybe just a firm finger.

Even perfect teeth are subject to repeated thermal stresses induced by hot and cold liquids and to repeated impacts due to the nature of eating and chewing food, forces sometimes supplemented by the grinding that takes place in some of us while we sleep. Such actions also can cause hairline cracks to start and grow over time, thus weakening a tooth and setting it up for breaking when it comes down upon an unexpected small piece of bone in a hamburger or part of a shell in an oyster stew. However started, progressed, and ended, this is the process of fatigue-crack initiation, growth, and fracture, something engineers have to take into account when designing bridges and roads over which traffic repeatedly bumps and bounces. Over seasons of freezing and thawing temperatures, and years of being chewed up by passing cars and trucks, our concrete and asphalt roads develop cracks that grow into gaps and into the cavities known as potholes.

One day, while eating lunch in Boston's Logan Airport, I sensed something different about one of my front teeth—it was loose. This was a capped tooth, a trophy of a teenage accident, which sat next to three other caps, further trophies of the same misadventure. I had always favored those teeth, avoiding biting into firm apples, breaking off hard pretzels, or pulling on tough taffy with them. When I returned home, I saw my dentist, who told me that the loose tooth was not salvageable. X-rays had revealed that the remnants of the real tooth, into which a stainless steel post had been inserted to anchor the cap, had broken in two along a line from the point at which the post entered it to where its root began. When I asked why this tooth and not its neighbors, my dentist showed me on the X-rays how the root canal into which the broken tooth's post had been inserted had been drilled at an angle, terminating closer to the root's boundary, and thereby from the outset leaving this assembly weaker than the others.

I still wondered, given the care I had taken with my teeth, why this tooth had given out exactly when it did. Eventually, after thinking about it, I came up with an answer that satisfied me. Some years earlier, when our son was young, I was installing a basketball backboard above our carport. The installation involved bolting the backboard frame to a heavy fascia board, and as I was tightening the bolt with a ratchet wrench, it slipped off the nut and struck me in the area just above my upper lip. The wound took several stitches and left a scar, but otherwise it seemed like just another badge of honor earned in the campaign of life. I was grateful that the wrench did not strike my capped teeth, but it never occurred to me that it had struck a blow through my skin and gum to the tooth's base. I suspect that that or another, forgotten trauma introduced some kind of nick or dent or small crack in the tooth, and it grew over time into a crack that was ready to give when I was eating lunch in the Boston airport.

The act of chewing, repeated perhaps hundreds of times daily, over the years had subjected my vulnerable tooth to hundreds of thousands of cycles of stress. This was the perfect environment for a crack to develop, grow, and let loose. The beating of our heart, which occurs about a hundred thousand times each day and a million times every ten days, can do a similar thing to the leads on an implanted cardiac device, such as a pacemaker or defibrillator. Flexing as they do with every beat, the tiny wires—perhaps inadvertently scratched by a surgical tool during the implant operation—can develop microscopic cracks that grow ever so slowly but deliberately to a critical size that leads to fracture and failure. The corrosive nature of human blood can aggravate fatigue and fracture problems in catheters and stents. The phenomenon of failure is ubiquitous.[2]

There is nothing new about things breaking. Sticks and stones have always broken bones, and our ancestors learned long ago how to fracture small rocks to make flint knives and arrowheads. Galileo used Renaissance experience with large stone obelisks and wooden ships inexplicably breaking spontaneously—though proportionately smaller ones did not—to motivate his research into the strength of materials. But it was the widespread industrial application of iron in the development of the railroads that brought the growth of cracks and the fracture of parts containing them to the widespread attention of engineers. When an iron axle, rail, wheel, beam, or bridge broke without warning, the result was often a spectacular accident, sometimes accompanied by loss of life. It was important to understand the ultimate cause of such failures in order to avoid them in building reliable railroad systems.[3]

Among the early researchers to study the fracture of railway axles was the Scottish civil engineer and physicist William John Macquorn Rankine. In 1843, while still in his early twenties, he presented a paper at a meeting of the Institution of Civil Engineers entitled, in the manner and style of its time, "On the causes of the unexpected breakage of the Journals of Railway Axles and on the means of preventing such accidents by observing the Law of Continuity in their Construction." It was a model of practical analysis within the context of a paucity of theory. In his paper, Rankine cited the prevailing hypothesis that the deterioration of metal over the course of repeated use occurred when "the fibrous texture of malleable iron assumes gradually a crystallized structure, which being weaker in a longitudinal direction, gives way under a shock that the same iron when in its fibrous state would have sustained without injury." Rankine admitted that this was a difficult hypothesis to prove, since the crystalline texture may have existed in the axle when it was new. Regardless, he proposed making axles not with the conventional abrupt steps in profile but with gradual changes in diameter, so that the metal's "fibre shall be continuous throughout." In other words, he recognized the deleterious effect of sharp changes in geometry. In fact, this is something we all tend to learn the hard way: an item we have clipped from the newspaper will tear easily at an inside corner. Good designs avoid such unfavorable geometry.[4]

In spite of Rankine's skepticism, the crystallization of iron under repeated loading prevailed for pretty much the rest of the nineteenth century as the conventional technical wisdom in explaining brittle fractures, meaning those that are not accompanied by any significant distortion of the material in the vicinity of the break. This is why a broken piece of china, if all the pieces can be found, can be glued back together into its original shape. Brittle fractures also result when a piece of glass or chalk is broken. When the separated parts are fitted together, only a hairline crack reveals the location of the failure site.

A famous brittle-fracture failure occurred in 1847 under the weight of a train passing over an important railroad bridge over the River Dee at Chester, England. This critical crossing in the famous London-to-Holyhead route provided a vital and strategic communications link between England and Ireland, via Wales and the Irish Sea. The Dee Bridge was a hybrid or composite design, with cast-iron sections trussed together with wrought-iron rods. Because cast iron is a brittle material that tends to fracture, beams made of it could not be more than thirty-five to sixty feet long, depending upon the engineer consulted. Thus the nearly 100-foot spans of the Dee were made up of three cast-iron girders arranged end-to-end, with the wrought-iron rods serving to keep the tandem assembly clamped together so that gaps could not open up between the beams. The wrought iron also served to keep the bridge from total collapse should the cast iron fracture. Bridges of a similar trussed design had been used on railroads since 1831, and because over the years they had provided reliable service they had come to be used for longer and longer spans and with lower and lower factors of safety—a natural evolutionary trend for structures of all kinds. The Dee spans were the longest of their kind ever built.[5]

In the immediate wake of the accident, which claimed five lives, the railway commissioners called for an investigation. They found that one of the trussed girders had fractured in a couple of locations. In an attempt to understand better what had happened, a series of tests was conducted by driving a locomotive over some remaining spans of the same design and noting how much they were depressed under it. The deflection was less than when the train was just standing on the girder, but as the train passed over the span there were noticeable vibrations set up in the supporting structure. Among the conclusions investigators reached was that under the action of heavy and repeated loads, "girders of cast iron suffer injury, and their strength becomes reduced." This was a primitive description of what has come to be known as metal fatigue.[6]

Because people were killed when the train carriages fell with the broken bridge span, an inquest was conducted. Painters who had worked on the bridge testified that it had indeed deflected considerably under passing trains, and the amount of deflection depended on the speed of the train. One painter used his ruler to measure deflections of up to four inches; another painter observed a deflection of five inches in a girder that later broke and was replaced. Testimony was also taken from several engineers, including Robert Stephenson, the bridge's designer. He had been responsible for many successful shorter bridges of the same type and insisted that the structure's design was not at fault. According to Stephenson, the accident on the Dee was initiated by a derailment, whereby the train struck the girder sideways and broke it. This explanation was contradicted by eyewitnesses.[7]

The coroner preempted jurors from finding Stephenson negligent but invited them to comment on the failed bridge's design.

The jury found that the "girder did not break from any lateral blow from the engine, tender or carriage, or van, or from any fault or defect in the masonry of the piers or abutments; but from its being made of a strength insufficient to bear the pressure of quick trains passing over it." The jury further asserted that "no girder bridge in future should be made of so brittle and treacherous a metal as cast iron alone. Even though trussed with wrought iron rods, [it] is not safe for quick or passenger trains." The jury feared that the hundred or so bridges similar in design to the Dee were "all unsafe." Its recommendation was that the government institute an inquiry to determine whether such bridges were safe. If they were not, they should all be condemned; if they were, the public should be so assured. A royal commission was established to look into the use of iron in railway structures. It conducted full-scale tests on cast-iron girders and confirmed that repeated loading did indeed decrease their strength. According to the commission's report, which was published in 1849, the broken beams showed "a peculiar crystalline fracture and loss of tenacity," which reinforced the idea that Rankine had expressed six years earlier.[8]

Strictly speaking, failure analyses are hypotheses heaped on hypotheses. How the bridge was designed, constructed, maintained, and used can have predisposed it to failure, and official and anecdotal records can provide grist for imaginative scenario mills. Generally speaking, the nature of the fracture surfaces of broken parts can provide valuable clues as to how a fracture proceeded, but they can be but one investigator's reading of the artifacts. Failure hypotheses can seldom be proven with absolute certainty because the structure no longer stands to be tested under the conditions hypothesized, and what evidence there might be is usually either incomplete or tainted. In the case of the Dee Bridge, parts of the fractured girder were lost in the river. Even when all fragments can be recovered, the fracture surfaces can be altered as the failure proceeds or mishandled in the process of recovery, thereby making conclusions drawn from them suspect. As a result, incidents like the Dee Bridge failure can be revisited and reinterpreted for years, decades, and even centuries. And we can continue to learn new lessons from them.

Sometimes circumstances preceding a failure call out for attention. Just before the Dee Bridge collapsed, five inches of ballast had been added to its roadway to keep embers spewed from passing locomotives from reaching the wooden deck and starting a fire there. The added dead weight of the ballast prompted the hypothesis that the fatal train was the load that broke the camel-like girder's back. Another hypothesis has been that the wooden planking, which was supported by the inner flange of the cast-iron beams, produced an asymmetrical load on the beam structure, causing it to twist under a passing train. This twisting may have initiated a structural instability, thus causing the girder to buckle and consequently fracture. This had been the prevailing explanation for decades.[9]

The most recent and novel hypothesis was advanced a century and a half after the fact by Peter Lewis and his colleague Colin Gagg of the Open University, coauthors of, among other publications, *Forensic Materials Engineering*, a book of case studies of product failure. According to Lewis and Gagg, the ultimate cause of the Dee Bridge disaster could be traced to an aesthetic design flourish cast into the beams. This feature introduced an area with abrupt changes in geometry and thus concentrations of stress that could have precipitated the growth of cracks from any tiny flaws in the casting. The idea that the seemingly innocuous desire to make a functional structure a bit more attractive can be the root cause of a failure is not far-fetched. The design of the infamous Tacoma Narrows Bridge was driven at least in part by the aesthetic goal of producing a long, slender-looking structure, a style that had become fashionable in the bridge design community during the 1930s. The slenderness of the deck of the Tacoma Narrows Bridge proved to be the structure's Achilles' heel.[10]

The scenario that Lewis and Gagg imagined for the Dee Bridge is as follows. The iron beams were cast with a detail known as a cavetto molding at the location where the vertical web met the horizontal bottom flange. Similar moldings were and are familiar finishing details where vertical walls meet horizontal ceilings in a house. Since carpenters would likely have been involved in making the forms for the cast-iron girders, they may well have introduced the detail as a finishing flourish, perhaps even thinking that it would produce not only a better-looking beam but also a better-performing one. Unfortunately, the sharp corners of the cavetto molding provided a site for concentrating stress (much as the steps in Rankine's axles did and much as a crevice provides an area for collecting dust). If the cast beam contained any flaw at all—such as a tiny void, nick, or other imperfection—it could serve as a nucleation site for a crack to grow a small amount with each passing train, thereby exhibiting the phenomenon of fatigue-crack growth. In time, the crack would have reached critical proportions, and the next time the beam was loaded by a train it would have given way. Thus Lewis and Gagg hypothesized that the Dee Bridge failed as a result of fatigue-crack growth initiated at the aesthetic flourish. Making things prettier should not come at the expense of making them stronger.[11]

Whether or not to blame the designer of the Dee Bridge for its collapse depends upon one's point of view regarding the extent of an engineer's responsibility to foresee failure. At the time the bridge was being built, the trussed-girder design had been in use for almost two decades, and Robert Stephenson himself had been using the form for over a decade. The focus of Stephenson and other engineers was on the weakness of cast iron in tension, which is induced in the underside of a bridge beam by the very loads it is

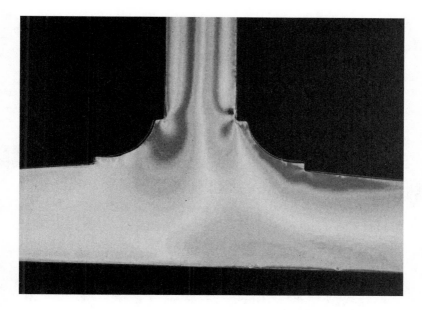

A transparent plastic model of the cross-section of a Dee Bridge girder shows the so-called cavetto detailing where the vertical web meets the horizontal flange. The flange is being inclined by an unsymmetrical loading representing the effect of a train crossing the bridge, and the patterns of light and dark show variations in the stresses induced. We now know that the sharp interior corner of the cavetto detail served to concentrate stresses, thereby promoting the growth of fatigue cracks from inadvertent imperfections introduced during the casting process. When a crack reached a critical size, the girder broke suddenly, causing the entire bridge span to collapse.

designed to support, and hence the use of the trussing. The phenomenon of what we now know as metal fatigue was not unknown at the time, but the mechanism by which it originated and progressed from an imperfection was incompletely understood. The successful—and generally fatigue-free—experience of engineers with the trussed-girder design had given them confidence in the form, and it was reasonable for them to believe that they had mastered it. Indeed, while the word *fatigue*, in the sense of human weariness, had its origins in the seventeenth century, its use in the context of the degradation of metal structural parts subjected to repeated loading and unloading gained currency only in the mid-nineteenth century. Engineers whose careers had matured in the "pre-fatigue" days might have been excused their ignorance, but no engineer could be forgiven for not considering the possibility of metal fatigue in moving parts after it was identified as the designer's nemesis.[12]

Railway accidents are virtually as old as the railways themselves. On September 15, 1830, the opening day of scheduled service on the Liverpool and Manchester line, a member of Parliament, who was stretching his legs while his southbound ceremonial train was taking on water, was fatally injured when he was struck by a northbound train traveling on the adjacent track.

As the early railway networks grew, accidents increased accordingly, with many being caused by stray animals wandering onto the tracks, drunken passengers falling onto them, and railway workers forgetting they were on a right of way. But as the rolling stock accumulated miles and wear, more and more accidents began to be attributable to equipment breaking. One particularly dreadful incident occurred in France in 1842. As part of the birthday celebration for King Louis Philippe, excursion trains carried well-wishers to Versailles. One of the last trains back to Paris consisted of two locomotives pulling seventeen carriages carrying a total of 768 passengers. Just after crossing a bridge on the route, the lead locomotive derailed and overturned, passengers were crushed in the momentum-fed crash, and the ensuing fire burned many victims beyond recognition. In all, fifty-six people died and a similar number were injured. According to experts who studied the accident, it was caused by the fracture of the axle of the lead locomotive.[13]

Railway axles and bridge beams were not the only structural

components fracturing spontaneously in the mid-nineteenth century. In 1854, the *Minutes of the Proceedings of the Institution of Civil Engineers* carried a paper by Frederick Braithwaite, "On the Fatigue and Consequent Fracture of Metals," which bore the running head, "Fatigue of Metals." In his paper, Braithwaite recognized that "fatigue may arise from a variety of causes, such as repeated strain, blows, concussions, jerks, torsion, or tension, &c." Among the many mysterious accidents that he believed to have been the result of metal fatigue were a cast-iron girder that broke under a beer vat that was repeatedly filled and emptied; leaking soldered joints between sections of copper pipe in another brewery; the repeated fracture of cast-iron cranks on a water pump at still another brewery. The paper was followed by discussion, in which Braithwaite credited the consulting engineer Joshua Field, who specialized in marine engines, with suggesting the term "fatigue" to characterize the "species of deterioration of metal" being reported upon. Among the discussants of Braithwaite's paper was Rankine, who found in it confirmation of ideas in his paper of a decade earlier. (The word "fatigue" in the sense of structural degradation is believed to have been coined by the French mechanician Jean-Victor Poncelet, who used it in his lectures at the military engineering school at Metz and wrote in 1839 that "the most perfect springs are, in time, susceptible to fatigue.")[14]

A century and a half after the Dee Bridge collapse, Peter Lewis and another colleague, Ken Reynolds, a forensic metallurgist, revisited another classic nineteenth-century bridge failure that over the years has been the subject of verse and other recountings, as well as of many investigations and reinvestigations. The story, briefly, is that the North British Railway wished to have a bridge built across the Tay River estuary at Dundee, Scotland, so that its coastal line could be more competitive with railroads that took an inland route, where they did not encounter wide waterways that required ferries to cross. The river at Dundee was very broad but relatively shallow, and so a long bridge with many spans resting on many piers had been proposed and designed by the engineer Thomas Bouch. Measuring two miles from approach to approach, the completed Tay Bridge was the longest in the world at the time of its opening at the end of May 1878.[15]

The bridge may have been long, but when built it was not especially daring in its individual truss spans, the longest being 245 feet at a time when such a length was not uncommon. The bridge's longest spans were its so-called high girders, which were set atop the piers to provide maximum clearance for tall-masted sailing ships passing below. The railroad trains were driven *through* the bridge structure at the high girders, rather than atop the trusswork as they were elsewhere on the bridge. The Tay Bridge carried rail traffic across the river for about a year and a half, but on the night of December 28, 1879, its high girders—which collectively amounted to over a half mile of the total bridge length—all collapsed as a train was running through them. All seventy-five people on board were killed.[16]

The Board of Trade appointed a tribunal to look into the causes of the accident and to assign responsibility for the tragedy. There were three members of the tribunal: William Henry Barlow, who was president of the Institution of Civil Engineers; William Yolland, an engineer who was Britain's chief inspector of railways; and Chairman Henry Cadogan Rothery, who was the government's wreck commissioner, but not an engineer. The voluminous testimony that resulted from the inquiry provides considerable insight into the design and operation of the bridge. Among the facts it reveals is that the cast-iron piers had contained numerous imperfections, that the girders and piers had not been securely tied down against the wind, and that the bridge had exhibited considerable vibrations when trains passed over it. However, the members of the court could not reach agreement about the absolute cause of the accident or the placement of blame.[17]

Barlow, Yolland, and Rothery agreed on the factors contributing to the failure. The three commissioners concluded that "the bridge was badly designed, badly constructed and badly maintained, and that its downfall was due to inherent defects in the structure, which must sooner or later have brought it down." However, the two engineers on the commission stopped short of specifying a definitive scenario, asserting that they had "no absolute knowledge of the mode in which the structure broke down." Chairman Rothery disagreed with his colleagues about the sequence of events leading to the collapse, and thus the commission's final report consisted of two parts, one reflecting Barlow and Yolland's conservative view and the other containing Rothery's more aggressive stance. Nevertheless, collectively they were not reluctant to place blame for the failure unambiguously on the bridge's engineer, then Sir Thomas Bouch: "For the faults of design he is entirely responsible. For those of construction, he is principally to blame in not having exercised that supervision over the work, which would have enabled him to detect and apply a remedy to them. And for the faults of maintenance he is also principally, if not entirely, to blame in having neglected to maintain such an inspection over the structure, as its character imperatively demanded." Bouch, who had been knighted upon the completion of the bridge, retreated from public view and died four months later, a ruined fifty-eight-year-old man.[18]

The conventional wisdom for over a century was that the high girders of the Tay had somehow been blown over while the train, whose carriages would have presented a large flat surface to the wind, was driving through them. There were reportedly gale-force winds on the night of the accident, but a contemporary photograph of the destroyed bridge clearly shows a number of tall smokestacks associated with Dundee jute mills standing undamaged in the background. According to the Beaufort scale, a force 9 wind should cause slight structural damage, such as broken chim-

The background in this photograph, taken in early 1880, provides evidence that at least some of the tall smokestacks in the city of Dundee, Scotland, survived the winds that blew on the December 1879 night that the high girders of the Tay Bridge—and the train crossing it at the time—fell into the river. That the smokestacks were still standing suggests that the winds were not so strong that they alone could have been responsible for the failure of the bridge.

ney pots, and several of these were reported to have occurred on the night of the Tay disaster. Such a wind exerts a mean pressure of about 7.7 pounds per square foot, and in absolute terms the pressure may reach the 10 pounds per square foot that was used in designing the bridge. (A factor of safety would have made the bridge capable of sustaining winds several times the design load.) Benjamin Baker, one of the expert witnesses for Bouch at the inquiry, surveyed damage in the area and concluded that wind pressures did not exceed 15 pounds per square foot, well below what should have been needed to topple the bridge.[19]

The photograph that incidentally showed the intact smokestacks was one of a series of images captured a week after the tragedy by a local professional photographer. The pictures had been ordered by the court of inquiry to provide a record of the accident scene and were used in the course of the proceedings to refresh the memories of witnesses. Among the photographs taken were those of the damaged towers that had bracketed the high-girder section of the bridge and the twelve piers in between that had supported it. In fact, most of the superstructure had gone down with the girders. The photographer captured the state of each pier and the surviving ironwork atop it from a variety of perspectives, including long shots taken from each adjacent pier and close-ups of debris taken from the pier itself.[20]

Peter Lewis first confirmed the existence of a photographic record of the Tay Bridge remains when he read the inquiry report. He knew that illustrations of the accident scene showing considerable detail had appeared in contemporary issues of *The Engineer,* and he knew that engravings were then commonly made from photographs. Now he wanted to find the original photographs, and he located a set of them in the Dundee City Library. From these he made high-resolution digital scans and studied the digitized images closely. What he found in the photos was evidence of broken pieces of cast iron that were identifiable as parts of the lugs that had been cast integrally with the columns designed to support the high girders. Lewis also found evidence in the photographs that the bolt holes were not drilled out but rather cast directly into the lug. Drilling the holes would have left them cylindrical, which would have provided relatively long parallel bearing surfaces for the bolt shafts; the casting process left slightly tapered holes, thereby setting up a condition in which the bolts fit loosely and the force they exerted on the lug was concentrated on a smaller surface, increasing the stresses in the lug. Elevated stresses were applied each time a train passed over the bridge, whose center of gravity at the high girders was literally high, thus making it quite top-heavy. The high stresses accelerated the growth of any cracks that might have developed in the lugs, and when the cracks reached a critical size the lugs failed. In other words, they failed after fatigue-crack growth. Some of the photographs confirmed this by showing characteristic patterns of incremental crack growth on the fracture surfaces.[21]

The lugs were designed to anchor the diagonal wrought-iron bracing located between the columns. The fact that the piers were littered with the remains of numerous broken lugs suggested that over time much of the bracing had effectively been lost in the support towers, rendering them more flexible and prone to a racking motion in a direction transverse to the rails. Such flexibility would likely have increased as the structure aged, with more and more lugs fracturing, resulting in no support for the associated tie rods. The vibrations set up by passing trains had driven the growth of fatigue cracks and the consequent failure of lugs, which in turn had allowed for larger-amplitude transverse motions and vibrations to occur and thus accelerate the deterioration of the structure. According to Lewis, the combination of a fast heavy train crossing the bridge in a strong wind must have caused the flexible towers and high girders to deflect sideways past the point of no return.[22]

The fatigue and fracture of metals were still incompletely understood in the 1870s, and even the word *fatigue* was still considered new; a monograph on long-span railway bridges published in 1873 put the verb *fatigue* in quotes. But even the great advances in theory and practice that have been made in the meantime have not eliminated the dangers associated with the repeated loading of a structure. In 1998, a German high-speed train accident that claimed one hundred lives was attributed to the fatigue failure of a wheel. In 2000, the derailment of a train in Great Britain was attributed to a fatigue failure that caused the steel track to break into hundreds of pieces. The crash of a fifty-year-old seaplane in 2005 was immediately suspected to have been caused by metal fatigue, and as recently as 2008 embarrassing problems with a cracked eyebar and a failed fix plagued the San Francisco–Oakland Bay Bridge.[23]

A bridge is not unlike a machine, in that each is an assemblage of parts that move relative to one another. As large as the bridge parts might be, their movements are typically relatively slight and not easily noticeable to the untrained, naked eye. Nevertheless, every bridge does flex whenever a vehicle passes over it, and for the Bay Bridge, this meant over a quarter million times a day. A bridge is also moved by larger forces. Gusts of wind can lift and drop it as turbulence does an airplane. Recall that in 1989 the Loma Prieta earthquake shook the Bay Bridge so violently that a section of its upper deck fell onto the lower roadway, killing a person. That incident ultimately led to the design of a replacement for the East Bay spans of the bridge, a project that continued for almost two decades.[24]

As noted, movements repeated over time can cause cracks to grow in steel components. If the progressive damage to a bridge is not detected and arrested before cracks become too large, collapse attributable to metal fatigue can occur. This is why inspections and proper preventative maintenance are so important. I once had a Volkswagen Beetle that I drove back and forth to work each day. The car was certainly nowhere near as important as the Bay Bridge, but, like all machines, the small automobile was made up of parts that moved and vibrated. In that car, I could feel the effects of wind gusts and potholes much more than I could in our heavier family station wagon.

One morning, sensing much greater noise and vibration than was normal for my Beetle, I stopped to see what was going on. When I opened the engine compartment, I saw that the generator was no longer firmly attached; the steel strap that clamped it to the engine block was cracked almost in two. It was a classic example of metal fatigue, and the strap had to be replaced. Not being near an auto supply store, I went into a close-by hardware store to get something to make a quick fix. The store did not carry generator clamps, so I bought some straps for securing an exhaust duct to a clothes dryer and used one on my car's generator and drove on to work. Afterward, I forgot all about my quick fix. Soon the noise returned, however, and with a vengeance; the lightweight make-shift clamp had broken after just two weeks of use. Since the stresses in its thinner section were much greater than in a standard clamp, the fatigue crack had grown much more quickly. I replaced the broken clamp with a new one and vowed to visit an auto parts store that weekend to get a proper replacement.

When a fatigue crack was found in a part of the Bay Bridge over a Labor Day weekend, a quick fix was devised and installed so that the heavily used bridge could reopen as soon as possible. The makeshift clamp was designed to supplement the cracked part, sort of like providing a pair of suspenders to help a steel belt hold up the bridge. Just two months after the fix, the five-ton pair of suspenders broke free and fell onto rush-hour traffic. Some vehicles were damaged, but fortunately no people were injured. This time the bridge remained closed for almost a week, while a more substantial repair was made. Plans were to inspect the new bridge fix daily. A subsequent study resulted in the design of damping devices that when installed would reduce vibrations and so also the risk of cracks developing from the small nicks and pits that were found to exist in the bridge's eyebars. Thus fatigue cracks, which plagued the earliest iron railway bridges, continue to threaten the safety of modern steel highway bridges. Understanding a problem does not necessarily solve it.[25]

We all occasionally take shortcuts maintaining, operating, and repairing our own machines, sometimes risking our bodies and our lives and the lives of those who rely upon us. But we do not expect those in charge of large pieces of public infrastructure to take shortcuts that place at risk the people who trust those machines to be in good working order. We should expect and believe that we deserve more from the custodians of our infrastructure. We should want them to think about failure in prospect rather than in retrospect.

And thinking in prospect means identifying all—not just the obvious—ways in which a system can fail. In the case of the Tay Bridge disaster, the poor design, poor construction, and poor maintenance of the structure certainly made each of those factors an easy target as a cause, and all three together may have seemed to be a damning combination. The engineer Bouch was working at a time when the fundamental mechanics of fatigue were not fully understood, but the existence of the *phenomenon* of fatigue was widely known and feared. There was no excuse for thinking that it was not something that could occur to a structure with high girders resting on spindly supports and subjected not only to the vagaries of the wind but also to the vibrations of passing trains. The debris that Lewis found scattered about the tops of the piers and that fed his suspicions that some of the lugs, at least, must have been broken well before the accident, should also have been clues to the initial investigatory team. Certainly the fractured lugs could have been accounted for in the catch-all of "poor mainte-nance," but it was the cause and effect of the fractures themselves that should have called out for more serious consideration. It should not have taken over a century to set the record straight.

SIX

The Old and the New

I have often been asked, "How long can a bridge last?" In other words, "How long after it is built will a bridge fail?" The answer ranges from negative days to months to decades on the one extreme and from centuries to millennia—and possibly even longer —on the other, depending on such diverse and interrelated factors as design, construction, materials, and maintenance, all of which are affected by the vagaries of economics, politics, corruption, weather, use, and luck. It also depends on what we mean by "fail." Examples are legion. The Quebec Bridge was not even completed when it collapsed into the St. Lawrence River in 1907. London's Millennium Bridge did get completed in 2000, but it stayed open only a couple of days before it had to be closed so investigators could study why it wobbled underfoot so much. The original Tacoma Narrows Bridge lasted only four months in 1940 before it was twisted apart by the wind. In 1967 a highway bridge that spanned the Ohio River for four decades collapsed suddenly under rush-hour traffic; an interstate highway bridge that for forty years had spanned the Mississippi at Minneapolis fell into that river in 2007. But much older bridges continue to function. The world-famous Brooklyn Bridge is now over a century and a quarter old. In England, the first iron bridge, completed in 1779, still carries pedestrians over the Severn River. And in southern France, the Pont du Gard stands as a monument to Roman engineering —two millennia after it was built. Engineers might and do argue that a properly constructed and maintained bridge can last indefinitely.

The thoroughness given to the initial design of a bridge is a principal factor in determining its lifetime. Among the chief decisions in designing a bridge is the material of which it will be made. Historically, timber and stone were used, and the latter is obviously more durable than the former. Who can imagine the Pont du Gard standing today if the aqueduct had been made of timber? So why has timber been used for any bridge? The answer is, mainly, convenience, speed, and economics. Generally speaking, it is easier, faster, and cheaper to erect a timber structure. Of course, timber is subject to rot and fire, necessitating regular rebuilding if the bridge is to remain functional. As a result, old wooden spans, such as covered bridges, which were covered precisely to protect them from the weather and so extend their lifetime, seldom have all of their original fabric in place.

A stone bridge took much more time and effort to build. In fact, the typical stone arch structure required first the erection of timber scaffolding, usually referred to as centering or falsework. This temporary structure was necessary to hold up the individual wedge-shaped stones, known as voussoirs, until all of them were in place and could push against one another to form a self-supporting arch. Once the stone arch structure was completed, the then-redundant centering was "struck" and collapsed, its timber to be retrieved for reuse, perhaps in the construction of an adjacent arch. Since traffic beneath a stone bridge was obstructed, if not totally blocked, while the construction process progressed, stone-arch bridges were usually not used across important waterways. Of course, once a carefully built stone bridge was in place, it could remain there for a long, long time.

With the introduction of iron and, eventually, steel, into bridge building, much shorter construction times were generally required and the result was a lighter structure. Of course, steel can corrode over time and hence must be protected. This is why steel bridges, especially those exposed to a corrosive environment such as salt-water spray or road salt, must be painted regularly. Concrete, which may be considered the successor to stone construction, is not immune to deterioration. It, too, is susceptible to damage by ocean spray or road salt, which can attack the structure's embedded reinforcing steel through cracks in the surface. The rusting steel can then push out against the concrete and lead to its spalling and the span's subsequent aesthetic and structural decline, if not outright failure.

No matter what the material of a bridge, among the important components of a responsible design should be the specification of a regular program of inspection and maintenance. One rule of thumb is that approximately 1.5 to 2 percent of the new-construction cost of a bridge should be budgeted annually for its maintenance. Thus, in as few as fifty years, cumulatively this could amount to spending in bridge-opening-year dollars as much on maintenance as the bridge cost to build in the first place. All too often, however, especially during fiscal pinches, shortsighted budgetary politics result in deferred maintenance. Postponing scheduled inspections or maintenance work, such as bridge painting, can result in disastrous consequences. When New York had its fiscal crisis in the 1970s, the city's historic bridges were among the victims. Only when they had deteriorated to the point of being unsafe were steps taken to catch up on deferred maintenance, which usually entailed rather expensive restoration work.[1]

The Waldo-Hancock Bridge deserves to be a classic case study of a structure that was once heralded as a masterpiece but that over time grew ugly and dangerous owing to corrosion, which compromised its strength and brought it progressively closer to failure. The bridge is named for the two Maine counties that it connects, thus carrying the coastal highway, U.S. Route 1, across the Penobscot River. With the completion in 1927 of the Carlton Bridge across the Kennebec River, about eighty miles south of the location of the Waldo-Hancock, crossing the Penobscot remained the last major gap in the coastal route. By 1929, there were four bills before the Maine legislature: three were to grant concessions to different companies to construct and operate a private bridge, and one was to establish a toll bridge owned and operated by the state. The last bill passed, but until the Waldo-Hancock Bridge was completed in 1931, motorists had to choose between driving an extra forty-five minutes to cross the river via the bridge at Bangor and relying on ferry service. The long detour could be faster than waiting for the boat.

The engineering firm selected to design the bridge between Waldo and Hancock counties was Robinson & Steinman, a consultancy based in New York. The association of the two engineers dated from 1920, when senior partner Holton Robinson had approached David Steinman about an international design competition for a bridge to connect the island of Santa Catarina to the mainland of southern Brazil. The innovative Florianópolis Bridge, which incorporated the structure's suspension chains into the stiffening truss of the main span, was the firm's first major commission. For the bridge across the Penobscot, David Steinman served as principal designer, and he produced a bridge whose main suspension cables were built up of twisted-wire strands that were hauled into place fully formed. This was a departure from the system that John Roebling had promoted and that is still employed in most long-span suspension bridge–building today, in which the main cables are built up of parallel steel wires individually carried back and forth between anchorages and across the towers. Steinman defended his design as more economical in cost and time for suspension bridges of no more than about 1,500 feet in total length.

The Waldo-Hancock bridge was to be exactly 1,500 feet between anchorages, with a central span of 800 feet.

The steel towers of the bridge were also a departure from the usual, which at the time were typically dominated visually by arches or crossed diagonals that served to stiffen the skyscraper-scale vertical structures. Steinman felt that at the Maine location, "the rigor of the natural rocky setting, the stern lines of adjacent Fort Knox and the background of colonial architecture in the neighboring town called for something simple." He thus employed a predominantly vertical- and horizontal-themed tower design that functioned structurally as what is known as a Vierendeel truss, which derives its strength and stiffness from the nearly rigid perpendicular rather than the diagonal arrangement of its components. The Golden Gate is among other large suspension bridges built in the 1930s that used such an alternative. It has incorporated into its upper-tower design, beneath the art deco façade, what is essentially a Vierendeel truss.[2]

The Waldo-Hancock Bridge was a model construction project, taking a total of only sixteen months (from August 1930 to November 1931) and coming in at about 70 percent of the original appropriation of $1.2 million. With the money saved, a replacement bridge was built between Verona Island, the eastern terminus of the Waldo-Hancock, and the neighboring town of Bucksport, which is located on the mainland. The rest of the left-over money was used to build roads in the vicinity. Dedication ceremonies for the Waldo-Hancock Bridge took place on June 11, 1932, and they included a report on bridge finances by the chief engineer of the State Highway Commission. David Steinman—the engineer-of-record representing the firm that was responsible for all surveys, design, and construction—turned over the completed bridge to the governor, William Tudor Gardiner, representing the state of Maine. Flags were raised atop the towers of the bridge as the assembly at nearby Fort Knox stood at attention. (The fort, which was built during border disputes with British Canada to protect upriver areas from the British navy, dates from 1844.) The exercises ended with the singing of *America* and a benediction, and were followed by band music and a baseball game.

Even before its formal dedication, the Waldo-Hancock Bridge had attracted favorable notice. In 1931 the American Institute of Steel Construction conferred on the structure its annual award of merit as Most Beautiful Steel Bridge. The first long-span suspension bridge in Maine, the Waldo-Hancock had long presented a striking prospect to motorists heading north on U.S. 1, as it came dramatically into view, and to boaters sailing up the Penobscot, who could appreciate it from a distance. (The view of the bridge for those driving south, after clearing a big bend in the road, was sudden and head on; the viewing time from a boat coming downriver was similarly cut short due to the river's sharp bend around Fort Knox.) The bridge was added to the National Register of Historic Places in 1985 and subsequently documented by the Historic American Engineering Record, with results deposited in the Library of Congress. In 2002, the bridge was named a National Historic Civil Engineering Landmark by the American Society of Civil Engineers.

Alas, even as the distinguished structure was being considered for landmark status, the wires in its cables were undetectedly corroding and snapping—the bridge was experiencing incremental failure. Such deterioration can long go unnoticed, since the cables of a suspension bridge are typically composed of a very large number of individual wires and the ensemble covered and painted for protection against the elements. In the case of the Waldo-Hancock, the first signs of trouble were discovered in 1992, when a limited section of one of the six-decades-old cables was unwrapped—a low section near mid-span, where intrusive water would naturally collect and promote corrosion—and thirteen wires inside were found to be broken. Since each cable had a total of 1,369 individual wires, the number of broken ones had a relatively small effect on the bridge's strength. The load that had been carried by the wires before they broke was naturally redistributed among the remaining intact wires, thereby increasing the load each had to carry by a small amount. But every bridge is built with a factor of safety, which is the ratio of the ultimate load it is capable of carrying divided by the actual load it was designed to carry. The Waldo-Hancock Bridge was built with a factor of safety of about 3, and the loss of the holding power of less than 1 percent of its wires did not diminish that appreciably. But the fact that any wires at all had failed was significant, and the condition of the cables was something to monitor.

It should be noted that in the United States, at least, the terms "factor of safety" and "safety factor" are often used interchangeably and, when a quantitative measure greater than 1 is associated with them, connote reserve strength. However, in some cultures, the term "safety factor" is used qualitatively and can have just the opposite connotation. Thus in Australian usage, "safety factor" means "an event or condition that increases safety risk." In other words, when a safety factor occurs it increases the likelihood of an accident. For example, in 2010, when an engine exploded on an Airbus A380 aircraft shortly after it took off from Singapore headed for Sydney, the event was described in an Australian Transport Safety Bureau report as a "safety factor." Even when dealing with technical matters, it is important always to be mindful of cultural differences in language and custom. Just as the word *failure* can take on different meanings in different contexts, so can *safety*. In any case, in a preliminary failure analysis the cause of the A380 engine failure was traced to an improperly machined turbine part that promoted the growth of a fatigue crack that reached critical size during the plane's takeoff from Singapore. The fatigue crack in question was obviously a safety factor.[3]

Whatever called, such conditions can also be present in station-

ary structures like bridges. In 2002, in conjunction with rehabilitation work being done on the Waldo-Hancock Bridge, the entire
north cable of the structure was unwrapped. To the surprise of
engineers, a considerably larger number of wires were found to be
broken than had been detected in 1992. A portion of the south cable was also inspected, and by comparing its condition with that
of ten years earlier, investigators could ascertain how fast the cables were deteriorating. (In comparison to the thirteen broken
wires found in 1992, eighty-seven were found ten years later.) According to calculations, in parts of the cable the factor of safety
had dropped to as low as 2.4, and projections indicated that it
could drop to the dangerously low level of 2.2 in four to six years.
Metal fatigue was partly to blame for the breaking wires. As was
the case with the repetitive loading of cast-iron beams of railroad
bridges and the wrought-iron axles of railroad cars, every time the
force in a cable wire was cycled, as it had been in the course of traffic crossing the Waldo-Hancock Bridge, tiny imperfections had
grown into small cracks which in turn had grown into larger ones
and thus reduced the residual strength of the wire. Eventually,
when the force in an individual cracked wire exceeded its reduced
strength, the wire snapped.

Large-truck traffic accelerates the process of fatigue-crack
growth, and there were concerns that all trucks might have to be
banned from the bridge. However, perhaps as a result of pressure
from truck drivers whose livelihood depended on access to the
crossing, officials restricted truck traffic rather than banning it
outright, and for a time they imposed a weight limit of twelve tons
on vehicles using the bridge. In the fall of 2002 the state posted
signs ordering heavy trucks to maintain a minimum distance of
500 feet between each other. This reduced the probability that the
main span of the bridge would experience an excessive number of
heavy trucks at the same time. Soon, the minimum distance allowed was extended to 800 feet, thus reducing further the chance

that too many large trucks would be on the bridge simultaneously and ensuring (if truck drivers complied with the spacing limit) that no more than two heavy trucks were on the main span at the same time. The Maine Department of Transportation also considered cracking down on overweight trucks. It was bad enough that the bridge had been designed when a ten-ton truck was considered large. Now, however, trucks weighing as much as fifty tons were allowed on the highway, and in reality some trucks weighed more than that. Repeatedly loading the bridge so far over design expectations had accelerated its deterioration, but completely replacing the main cables was not considered practical.[4]

By the summer of 2003, it was clear that, even with restrictions on truck traffic, the Waldo-Hancock Bridge had a limited useful-life expectancy, perhaps as few as four-to-six years. Since that was also the amount of time it might take to design and build a new bridge, it was clear that necessary decisions had to be made and decisive action taken very quickly if the region was to continue to enjoy a convenient and fully functional fixed crossing of the Penobscot River close to its mouth. By putting the project on the fast track, state planners thought a new bridge might be ready for traffic in perhaps three years. The Maine Department of Transportation worked with the Figg Engineering Group design team and the construction joint venture undertaken by the firm of Cianbro, which works jobs all over North America, and the relatively small Maine-based Reed & Reed, which does work principally in northern New England.

Among the most important decisions in designing a new bridge are where to locate it and what kind of bridge to build. Obviously, the original Waldo-Hancock was already in place, and so its exact location (perhaps the ideal one for a bridge, at least at one time) was not available for a new structure. Locating a new bridge right next to the old makes sense because then only minimal realignment of existing approach roads is necessary. However, the existing northbound approach to the Waldo-Hancock hugged the high riverbank and so had required drivers to make a sharp right-angle turn on entering or exiting the bridge. Locating a new bridge a bit downriver from the old and blasting a new stretch of road out of nearby rock presented an opportunity to incorporate a more gradual approach curve. As for the kind of bridge to build, the lay of the land and navigation requirements immediately called to mind three obvious choices: another suspension bridge, an arch bridge, or a cable-stayed bridge.

The citizens of the adjacent counties naturally had strong, if not sentimental, feelings about the familiar Waldo-Hancock Bridge, and they worried about how a new span would affect traffic patterns and the aesthetics of the area. The Waldo-Hancock's designer, David Steinman, who in addition to designing bridges would publish several books of poetry in his lifetime, had prided himself on seeing beauty in bridges and on making aesthetically pleasing ones that harmonized with their environment. Thus, as we have seen, in designing the Waldo-Hancock towers as Vierendeel trusses, he had eschewed soft curves and chosen instead the strong vertical and horizontal lines that echoed the hard edges of the nearby granite cliffs and ledges that characterize coastal Maine in the region. His finishing touch on the bridge was to have it painted green, which made it fit nicely with the pine trees and other evergreen foliage of the area. Steinman saw his bridges in context. (Today, most major steel bridges in Maine are painted the same pleasant green color.)[5]

With the decision made to replace the old, familiar structure with a new one, county commissioners, environmental activists, preservationists, and other interested citizens wanted to be informed about, watchful of, and involved in the choices that would affect the sense of place and quality of life in the environs. Many of the stakeholders wished to see the historically significant bridge remain in place, most likely as part of a pedestrian and cyclist path, and so they expressed concerns that the new structure might ruin the famous picture-postcard scene by upstaging the old. They also wanted the design of the new bridge to be compatible with the existing one. Then, of course, there was the issue of the cost of building the new bridge and restoring the old one. Should such concerns lead to legal issues, not only would the new bridge be delayed but also traffic on the rapidly deteriorating old structure might have to be restricted even more than it had been. To forestall such developments, planners established a bridge project advisory committee to facilitate communication between citizens and the state, especially as embodied in the Department of Transportation.[6]

When it became clear that the accelerating deterioration in the cables of the existing bridge might force it to be closed for safety reasons long before the new one was completed, the transportation department came forth with a plan to prevent that from happening. To avoid a total ban on truck traffic for the duration of construction of the new bridge, they proposed adding supplementary cables to the Waldo-Hancock above the old ones, thereby enabling some of the weight of the existing structure and its traffic to be taken off the original but weakened cables. This unusual measure—a unique procedure in the United States—naturally added to the cost of the entire project, but it also allayed the concerns of local truck operators and bought insurance time to allow for any unforeseen delays in the design and construction of the new bridge. The added cables were expected to carry about 50 percent of the dead load of the structure and increase the factor of safety from a low of 1.8 to 3.2, assuming that the bridge was carrying trucks weighing no more than forty tons. In late 2005, after the supplementary cables had proven themselves over two winters and the completion of the new bridge was in sight, trucks weighing as much as fifty tons were once again allowed to cross the existing bridge.[7]

In the meantime, as local and regional newspapers carried stories of progress on the new bridge, the once-clean lines of Steinman's Waldo-Hancock Bridge were cluttered with its life-support systems, and comments on the status of the landmark were generally relegated to all but footnotes to the celebratory news of the new. The rising bridge was expected to be a signature structure for the area, and where once there was talk of costly rehabilitation of the old bridge to make it into a pedestrian way, increasingly locals seemed resigned to its more likely fate—demolition—at a cost estimated to be as much as $15 million. This would be a sad fate for a National Historic Civil Engineering Landmark whose plaque had not yet even been mounted owing to the disruption caused by the deterioration of its cables.

Once it had been decided that a cable-stayed bridge would replace the old suspension bridge, it was time to make the partly aesthetic decision about the shape of its towers and how its cables would be arranged, and the fully technical decision about how they would be installed, anchored, and protected against the elements. The stay-cables might be thought of as the weak links in a cable-stayed bridge. They may be subject not only to vibrations that can lead to fatigue failure but also to corrosion in a harsh and aggressive environment, such as exists on the Maine coast and had been a factor in the deterioration of the Waldo-Hancock Bridge. The manner in which the new bridge's cables were to be installed and protected over the Penobscot was a significant innovation associated with the project. The design team decided that instead of each cable originating at a certain place on a tower and terminating at a particular place on the bridge deck, as is the case with most cable-stayed bridges, each cable would originate in the deck on one side of a tower, pass through a specialized conduit in it, and terminate at a corresponding location in the deck on the other side of that tower.[8]

The details of the bridge's cable system reflect the way that individual design decisions are made in response to and in anticipation of failure. In a typical cable-stayed bridge, matched pairs of cables push down on the tower but do not exert any significant net transverse or bending forces, which could tend to open up cracks in the concrete structure. In the new bridge, engineers employed a patented technique whereby each of the individual strands (essentially lengths of wire rope) contained in a full cable is anchored individually in the bridge deck and passes through a separate stainless-steel tube arranged in a cradle embedded in the pylon (the term often used for a cable-stayed bridge tower). In this way, a corroded, broken, or otherwise ineffective strand can be removed and replaced without disturbing the others—or in any significant way the traffic on the bridge. This system also allows for experimental strands, perhaps made of new but untested materials, to be used in limited quantities in an actual bridge, as a way of testing them—perhaps to failure—under actual bridge loading conditions. Since there are dozens of strands in a single cable, these untried strands present no safety hazard to the structure. The Penobscot cable-stayed bridge was only the second one in the world to incorporate this kind of cable system, the first being the Veterans' Glass City Skyway Bridge in Toledo, Ohio.[9]

To protect the stay-cables against corrosion in the harsh saltwater environment that contributed to the deterioration of the suspension cables of the Waldo-Hancock Bridge, each strand of the new bridge's cables was treated with a void-filling epoxy coating that keeps any water or other harmful elements that might be present away from the steel. The collection of strands that makes up each cable is also encased in an outer sheath made of hard white plastic (high-density polyethylene, or HDPE) designed to keep water, snow, ice, and other undesirable environmental agents out. To ensure that this system functions properly, each HDPE tube is filled with nitrogen gas under pressure that is constantly monitored by computer. The gas not only excludes any oxygen that might promote corrosion but also serves as a leak detector. If a pressure drop develops, it will be immediately detected and can be traced to any cracks, punctures, or other defects that have opened up in the sheathing.[10]

The groundbreaking for the new bridge had taken place in December 2003. At the time, construction was expected to be completed in two years at an estimated cost of about $50 million. By the fall of 2005, the completion date had slipped by about a year and the cost had risen to $84 million, one hundred times the cost of the original bridge in the dollars of its time. Among the reasons given for the increase were the higher prices of construction materials. In early 2006, officials announced that the new structure would be called the Downeast Gateway Bridge, but local politics led to a reconsideration, and the name the legislature finally approved was Penobscot Narrows Bridge and Observatory—the latter part of the name referring to the 420-foot-high observation deck that is located atop the west tower. One state senator did not think all the fuss over the name was worth it. She believed that local people would refer to the structure as the "new bridge" or as the Bucksport bridge, as they had called the Waldo-Hancock.[11]

Perhaps the most visible innovation associated with the Penobscot Narrows Bridge and Observatory is reflected in the last word of its elongated name. To see the big picture is a near-universal desire; to stand where one can get a bird's-eye view of the surrounding area seems to be an enjoyable experience to all but the acrophobic. Who has climbed, driven, or taken a funicular or other type of inclined railway or cable car up a hill or mountain and not paused to admire the view from the top? Hot air balloons gave eighteenth-century adventurers a perspective from even higher altitudes. Tourists have long sought out high vantage points, and tourist attractions have exploited this desire. The Washington Monument, whose form is mimicked in the towers of the new Penobscot bridge, became famous for its observation deck, as did the

The steel tower of a badly corroded Waldo-Hancock Bridge is dwarfed by the concrete pylon of the new Penobscot Narrows Bridge and Observatory. The adjacent suspension and cable-stayed spans, respectively, were each designed to carry U.S. Route 1 across Maine's Penobscot River, and the differences in their materials and form demonstrate that dramatically different solutions can exist for what is essentially the same engineering problem.

Statue of Liberty. The Eiffel Tower, the first structure to reach the long-sought-for height of 300 meters, owed some of its renown to its famous viewing platforms. Among the attractions of the 250-foot-diameter Ferris wheel erected for the 1893 Columbian Exposition in Chicago was the overview that it gave its riders from what might be thought of as mobile observation decks.[12]

With the growing availability of steel in the late nineteenth century, office buildings began to rise to unprecedented heights, which led them to be called skyscrapers. It became natural to cap such structures off with viewing levels. The Empire State Building, which held the record for the world's tallest building for four decades, has perhaps one of the most famous observation decks. A city's tallest buildings often vie with one another through the height and panoramic views of their uppermost floors. Incorporating tourist-attracting viewing levels into tall buildings has become more or less an expected feature of super-tall structures. The Sears (now the Willis) Tower and the John Hancock Tower provide excellent views of Chicagoland. Boston's John Hancock Building provides views of that city and its environs. Similar things may be said of observation decks in buildings around the world. Before the destruction of the World Trade Center in 2001, the topmost floor and roof of its north tower afforded spectacular views of New York Harbor—and the Statue of Liberty. Now, of course, concerns about terrorist attacks have limited access and increased scrutiny of those adventurers seeking to climb or ride up in a confined space to look down upon scenes that can seem to have no boundaries. In spite of this, it was just a matter of time before observatories appeared atop bridges, even though the idea had been around for quite a long time. Indeed, the observatory atop the Penobscot River crossing is said to be the first in the Western Hemisphere.[13]

An observatory—and its attendant complications—was probably far from the minds of those initially charged with designing a bridge to replace the rapidly deteriorating Waldo-Hancock. But that changed when researchers at the Department of Transportation, while looking into historic shapes that might be appropriate for the towers of the new bridge, found that granite from nearby Mt. Waldo had been incorporated into the Washington Monument. The idea understandably followed to model the bridge's pylons after the obelisk-shaped structure in the nation's capital. From this decision naturally arose the possibility of incorporating an observatory at the top of one of the towering pylons. This concept appealed to area business and civic leaders, who now saw the new bridge also as a tourist attraction, with its concomitant economic benefits. Since the support of the local community was important to the Department of Transportation, lest opposition to the bridge design delay progress on its construction, the observatory became an important and permanent part of the project.

To me, some of the most interesting things to see from the new bridge's observatory are located almost directly below it. Just to the north sits the rusting Waldo-Hancock, its towers dwarfed by the pylons of the new structure. When I first visited the bridge in 2007, the old crossing's supplementary cables were still in place, but its deck was empty of vehicles since the bridge had been closed to traffic on the next-to-last day of the previous year. A 1915 Ford was the last vehicle to cross the obsolete and deficient structure. The old bridge approaches were still in place, but they led to a roadway to which access was unceremoniously blocked by closed chain-link gates posted with "No Entry" signs. (When I returned to the bridge three years later, the only thing that appeared to have changed was that the old bridge had acquired more rust.)[14]

No doubt some preservationists would still like to have seen the old bridge restored to its as-built condition (including the removal of the supplementary cables) and returned to service as a pedestrian and bikeway crossing, but that would be an expensive proposition. The state's plan was to disassemble the old bridge, most likely removing parts of it in the opposite order to which they were erected, but with the aid of cutting torches and perhaps in the final stages explosives. But even demolition would cost millions of dollars, and so without any secure funding they did not have a firm timetable. There was some indication that connected with any removal contract would be the construction of a pedestrian walkway beneath one of the cantilevered sections of the new bridge's roadway. The walkway would be hung from the wing-like edge of the roadway and provide a means of separating pedestrians from vehicles. Until that might happen, a kind of shoulder or breakdown lane was to serve for pedestrians and cyclists, though they might be separated from vehicles only by the white line marking the traffic lane.[15]

From the observation deck, the visitor can follow the lines of cables that emerge from the pylon directly below in two vertical rows and converge into the single horizontal row of them anchored in the deck. This anchoring is done through the cable sheaths incorporated into the elevated median that forms the backbone of the roadway. From the observatory it is also strikingly clear that the new bridge has only a single traffic lane in each direction, thus appearing to present a potential bottleneck. When someone asked the observation-tower guide why the bridge was not built with at least two lanes in each direction, he cited "money" as the reason. He did offer that each side of the bridge deck was capable of carrying two lanes of traffic if need be—though with only ten-foot-wide lanes and no shoulders—but what would that leave for pedestrians and cyclists? According to a Department of Transportation technician, the bridge was designed with only a single lane in each direction because the historic nearby towns of Searsport to the south and Bucksport to the north could not accommodate an expansion of Route 1 to four lanes. Increasing the capacity of the bridge could thus have led to bottlenecks elsewhere.[16]

Money, or the lack of it, is often stated to be a major reason that something is done—or is not done—occasionally leading to failures that are not structural or catastrophic but that are blemishes on the design nevertheless. Perhaps it was money, or perhaps it was the rushed schedule to complete the bridge, but some design aspects of the new structure are disappointing functionally and aesthetically. The way the lanes are marked, there is no appreciable shoulder between the traffic and the massive cables anchored atop what is suggestive of an unbroken line of the kind of concrete Jersey barriers that one encounters on turnpikes in more densely populated states. But these in Maine are topped by a line of thick cable sheaths all aimed like howitzers up at the towers toward which the taut trajectories of the cables lead. According to the lane markings, all traffic is expected to drive close to the elevated line of cables. This scheme may have the structural advantage of minimizing any twisting of the roadway, but for some drivers and passengers it will mean being uncomfortably close to what might seem an encroaching center barrier. As a result, drivers and their passengers miss the sense of openness they might have wished for in a bridge in rural Maine; they experience, instead, a sense of oppressiveness, at least on the driver's side. Had the unused narrow shoulder to the traffic's right and the traffic lane itself been switched, the crossing might appear to drivers to be much more sensible and pleasant. But less evident structural considerations often govern a design, not to mention the complication of getting pedestrians and cyclists inside the lane of vehicle traffic, which would then block their view of the river. Nevertheless, from the elevated observation deck it is clear that many cars tend to keep their distance from the oppressive median and ride with their right wheels over the line, potentially encroaching upon walkers and bicycle riders. This aspect of the bridge's design might be said to be a functional failure—and a potentially dangerous one.

What might be termed an aesthetic failure, fault, or distraction, at least to this self-appointed structural critic, is the lack of continuity, balance, and symmetry in the towers. One noticeable feature of the towers is that the pylons above the roadway are narrower in the bridge's transverse direction than they are in the direction of traffic flow. The narrowness of the pylons was perhaps dictated by the need to allow space on the deck for the traffic lanes; more likely the thickness of the pylons in the bridge's longitudinal direction was meant to accommodate the system of cable conduits. When viewed obliquely from afar, however, the mismatch between the shape of the rectangular pylon above the deck and that of the apparently square pier below the deck is even more distracting. This may be an unfortunate consequence of the fast-track design-construction process, which may have allowed the foundations and piers to be built before the bridge type and superstructure design were finalized.

Another distracting feature of the bridge towers occurs near their top. The pylon containing the observatory is encircled by three clearly visible levels of plate-glass windows that make this part of the structure literally transparent. When my wife and I viewed the bridge from Fort Knox, we could see three distinct levels of people silhouetted against the sky on the far side. The delicacy of the glass-enclosed tower top is striking, and in certain light the fenestration gives the impression of three reflective bands just below the pyramidal top. However, the east tower, having no observatory, was not given a similar, balancing crowning treatment. Instead, the granite structure of it rises above the cables unrelieved or unaccented by anything until it begins to taper into its pyramidal shape, which is punctuated at the top by the red aircraft-warning beacon that sits at the apex.

The arrangement of cables on this bridge is also distracting. Issuing as they do from each saddle level in pairs of cables about ten feet apart horizontally and terminating along the single centerline of the deck, they create a three-dimensional funicular geometry that takes on a variety of confusing and dissonant patterns when viewed from different vantage points. When we looked at the bridge from the fort, the west approach-span cable arrangements appeared to have a line of intersection nodes about half way up. Other sets had a moiré-like effect that moved with our point of view. Cable arrangement is normally one of the aesthetic strengths of a cable-stayed bridge, but in the case of the new Penobscot span its complex set of interactions with depth and length seemed, on our first visit at least, to be confusing, distracting, and discordant.

It is easy to criticize any structure; it is hard to remember all of the constraints (including the political and the economical) that its designers, engineers, and constructors had to work under to produce a safe and effective piece of infrastructure. The new Maine bridge was opened to traffic just three years after groundbreaking took place. It had taken about six months to complete the foundations, and another six months to erect the piers on them. Had this phase of construction beneath the roadway level of the bridge not been done on the fast track, the bridge might still have been under construction four years after the groundbreaking. And that might have been longer than the Waldo-Hancock Bridge could have withstood the traffic it was being asked to continue to carry.

Valid points of criticism do not necessarily mean that a bridge design is a total failure or that it is not safe or suitable or sensible under the circumstances of its design and construction. Even with what may be perceived to be its forgivable flaws, the new bridge across the Penobscot River is a striking addition to the region's infrastructure. The Department of Transportation and firms involved should be praised for creating a striking bridge and observatory designed and built under the extremely trying conditions of urgency, economy, political pressures, and Maine winters. It is nice to think that the new bridge might last forever, but that would be wishful thinking.

As I write this, the Waldo-Hancock Bridge and the Penobscot

Narrows Bridge and Observatory still stand side by side, presenting a striking contrast of what might have been and what will likely be. When backlit by sunrise or sunset, the structures stand in silhouette, which suppresses all signs of physical age. But when the sun is overhead, it reveals the Waldo-Hancock to be a rusting hulk moored across the river, its deck hanging by what from a distance looks like threads. What should have been a proud historic structure, freshly repainted its Maine-bridge green, gives only a hint of that hue. Rust encroaches on it everywhere, and it is likely only a matter of time before the structure falls to the spreading corrosion or to the cutting torch. The latter will probably not happen soon, since money is tight and every year the bridge stands the more it will cost to take it down. As much as the old bridge is an eyesore, it is so much overshadowed by the height and brightness of the new one that it has become increasingly ignored by all but the sentimental historians of the region and the state Department of Transportation, which knows that it has a liability on its hands. Perhaps it would be best if the Waldo-Hancock were allowed to stand, neglected and rusting as it is. Its steady deterioration may not be pretty, but it would be a constant reminder to all who were willing to focus on it that the same fate might befall the newer structure some day—if it were not properly maintained.

SEVEN

Searching for a Cause

Some structures deteriorate slowly and quietly from within, their progress toward total failure as hidden from view as their buried foundations. But if the flawed system is not discovered in time to do anything precautionary or corrective, we can be taken by surprise by an abrupt and incontrovertible failure, such as the collapse of a bridge carrying rush-hour traffic; the explosion of a space shuttle one minute into a mission on a cold winter morning; or the sudden flooding of a city when its aging levees do not hold. The effect of such occurrences is immediately obvious, but the root causes can be maddeningly elusive. Investigations can take months, if not years, and controversy over the resulting official failure analysis can be as lingering as the event was sudden. More often than not, investigators identify a number of interrelated causative factors, while dismissing other, rumored ones as unsubstantiated or irrelevant, if they mention them at all. Assigning responsibility for the failure can be equally elusive. Sometimes names are named, but many times they are not. The people who may be responsible for failures tend to remain as anonymous as they were when the ill-fated system was being designed and built.

When an accident occurs, especially when it happens suddenly and progresses quickly, confusion understandably reigns among the people who are injured and trapped in the wreckage; the people dazed but able to walk away; the passersby, the onlookers, and other witnesses to the event; the first responders and the rescuers; the police and firemen; the reporters and other members of the news media; the family and friends of the victims and feared victims; the owners of whatever it was that failed; the designers of the failed structure; the ghoulish souvenir hunters; the curious; the insurance adjusters; the lawyers. Confusion feeds rumors, and the sequence of events can become less and less clear. It is often a day or so before the official accident investigators arrive on the scene, and members of any commission appointed to study the accident may not see it for weeks or months after the event.

It is no wonder that so many contradictory stories tend to accompany a tragedy. Different people observing the same occurrence from different perspectives will naturally see things differently. What is in one person's forefront is in another's background. One person stares while another blinks. One is looking up when another is looking down. One is walking and looking sideways while another is stationary but daydreaming. Some see with their eyes but others hear with their ears. They all witness the same thing but use different words to communicate their experience. A forensic investigation has to sort through all of this—and more—to try to come up with what really happened and why.

In the case of the collapse of the Kansas City Hyatt Regency Hotel's elevated walkways on July 17, 1981, an accident that claimed the lives of 114 people and injured many more, the mechanical cause of the failure was correctly reported on within days. The *Kansas City Star* had hired a consulting engineer to help interpret the accident, and he easily deciphered the evidence on the scene, photographs and drawings of which were published on the front page of the July 21 edition of the newspaper—just four days after the tragedy. One photo showed clearly a bent hanger rod dangling from the atrium roof, washer and nut still attached on the free end but restraining nothing. This obviously indicated that the box beam the assembly had supported had somehow passed over the end of the rod (and washer and nut) and separated from it. On the lobby floor, among all the wreckage, was clear evidence of how that separation took place: the steel surrounding the hole in the box beam through which the support rod once passed was deformed as if it was pulled down over the nut-and-washer assembly by some great force. In the ensuing months, the National Bureau of Standards conducted physical tests on replica box-beam connections and verified that the forces necessary to accomplish such a drastic separation were indeed within the range of those acting on the actual walkways on the fatal day.[1]

The tests, employing machines not nearly as large as the three-million-pound apparatus in Talbot Lab, did confirm the failure hypothesis, and additional analysis made it clear that the walkway connections even as originally designed should have been made stronger. But being underdesigned according to building standards does not necessarily mean that a connection detail will fail. In fact, had the single support rod of the original design not been replaced with two, thereby doubling the force between the box beam and the support washer—and halving the factor of safety from about 2 to close to 1—the walkways would likely still be standing today. The fact that the walkways did not satisfy the Kansas City building code was not the direct cause of the failure; the fact that the design detail of one rod was changed to two was. But not all accidents are so quickly analyzed nor so easily explained.

The bridge across the Ohio River between Point Pleasant, West Virginia, and Gallipolis, Ohio, which had served the region well since 1926, collapsed catastrophically ten days before Christmas in 1967, killing forty-six people and injuring dozens more. Officially named Point Pleasant Bridge, the structure was commonly referred to by its sobriquet, Silver Bridge. In fact, this nickname was often preceded by the definite article, because the structure was believed to have been the first bridge in America on which aluminum paint had been used. Just as with the iron-and-glass Crystal Palace that housed the Great Exhibition in London in 1851, what was once the silver-colored bridge's catchy unofficial name became its virtually official one. Although it was a critical link on U.S. Route 35 connecting the state capitals of Charleston and Columbus, outside of that region of the country and outside the bridge-building community, the Point Pleasant Bridge was not a widely known structure. But the implications of the failure and its cause would be profound for the bridge-building community and for the entire nation. The official investigation of the collapse was conducted by what was then the newly constituted National Transportation Safety Board, which is the agency that now is perhaps best known for its investigations of airplane accidents.[2]

Silver Bridge was a suspension bridge of a different kind. Instead of its roadway being suspended from large and heavy steel-wire cables that sagged in a parabolic arc between towers, as was characteristic of virtually all American suspension bridges built since the mid-nineteenth century, the deck of this bridge was hung from long-link steel chains. That in itself was not novel, for iron chains had been used in many an early-nineteenth-century suspension bridge, including the famous Menai Strait Bridge, the critical link in the strategic London-to-Holyhead Road. But the use of suspension chains on the scale of Silver Bridge was highly unusual in America. What was really different about its design was that its suspension chains doubled as the top chord of the trusses that stiffened the roadway, a system that had not previously been used in the United States. Robinson & Steinman had introduced the concept in Brazil, but even there it was only the 1,120-foot main span that incorporated the detail. In addition to the trusses stiffening the 700-foot-long main span of Silver Bridge, the side-span trusses also had the Florianópolis feature, making for a total of 1,460 feet of common links serving as both suspension chains and the top chords of the trusses.[3]

The original design for the bridge between Point Pleasant and Gallipolis was more conventional than the structure that was eventually built. The Baltimore firm of J. E. Greiner Company had determined that by employing the by-then familiar combination of steel-wire cables and a distinct stiffening truss, the bridge could be built for an estimated cost of $825,000. However, when the construction job was advertised, bidding contractors were given the opportunity to propose an alternative design that could be erected for under $800,000. If that could be done, the contractor would receive half of the amount saved. The winning bid was submitted by the American Bridge Company, which had built the Florianópolis Bridge and which put forth the design employing eyebar suspension chains, some of whose links would double as truss members. Parallel pairs of eyebars, each of which was to be two inches thick, twelve inches wide, and as long as about fifty feet, would be linked together bicycle-chain style with steel pins to form the main part of the suspension system. And, as in the Florianópolis Bridge, the steel to be used in the Silver Bridge eyebars was to be— also for the first time in the United States—heat-treated to produce a stronger material. Because chain links made of such steel could carry more load relative to their own weight, the bridge itself would be a lighter and thus less costly structure. However, there was one notable difference between the chains of Silver Bridge and those of the Florianópolis. The latter had four eyebars per link, whereas the former had only two. This would prove to be a very significant weakness in the American design.[4]

Unusual design features of Silver Bridge did not end with its eyebars. The towers over which the chains passed were not rigidly fixed on their piers, as was common, but were designed to be able to rock freely back and forth on them in response to slight changes

Silver Bridge, more formally known as Point Pleasant Bridge, after the town in West Virginia that it served, was of an unconventional design, in that some of the large steel links in its eyebar suspension chains also served as upper-chord members of the structure's stiffening trusses. The bridge collapsed suddenly in December 1967, a failure that was ultimately attributed in part to corrosion-assisted fatigue-crack growth that proceeded without detection. The incident led to the requirement that all U.S. highway bridges be inspected on a regular basis.

in cable pull. In contrast, the main suspension cables, whether made up of chain links or individual steel wires, had to be solidly anchored in some way, so that the great tension in them could be resisted and their ends could be kept in the proper position. Thus regardless of how the suspension cables or chains may have moved with the towers, their ends had somehow to be fixed securely to the ground. The preferred way of doing this was to embed the terminal lengths in bedrock, but if this was not close enough to the surface at the bridge location an alternative had to be found. The typical alternative was to construct massive masonry or concrete monoliths whose sheer weight could resist the pull, which in the case of Silver Bridge was 4.5 million pounds per chain. In lieu of conventional anchorages, those of Silver Bridge consisted of "reinforced concrete troughs that were filled with soil and concrete" and were sitting on reinforced-concrete piles. The troughs were 200 feet long, and when paved over its top, each anchorage doubled as the surface of that portion of the approach road.[5]

The 1920s was a time of great innovation and experimentation in suspension-bridge design and construction. Simultaneous with the design of Silver Bridge, that of a twin-sister structure was being prepared by John Greiner's firm. This bridge, located ninety miles upstream, where it connected St. Marys, West Virginia, and Newport, Ohio, would open in 1928 but be disassembled in 1971, when no assurance could be found that it would not suffer the same fate as Silver Bridge. (The Florianópolis Bridge was not disassembled after the Ohio River accident because its four eyebars per chain link provided a greater degree of redundancy and thus an ability of the chain to maintain its basic geometry even should one of its eyebars fail.)[6]

Three other notable eyebar chain suspension bridges, but without integral trusses, were constructed between 1924 and 1928 in Pittsburgh, Pennsylvania. These three—the Sixth, Seventh, and Ninth Street bridges, formally named for Pittsburgh notables Roberto Clemente, Andy Warhol, and Rachel Carson, respectively— are collectively known as the Three Sisters. They have the added distinction of being self-anchored suspension bridges, which means that their chains terminate at the ends of the stiffening girders that support the roadway. This alternative was employed because Pittsburgh's City Art Commission expected "aesthetic" structures to be designed, and engineers with the city's public works department did not think massive anchorages were compatible with that nontechnical constraint. However, employing the self-anchoring principle meant that the stiffening girders had to be sufficiently deep and heavy to be able to resist the great compressive forces imposed upon them by the chains. While this may have been an acceptable appearance in the 1920s, the bridges soon looked malproportioned because structural aesthetic standards, which follow fads and fashions the way clothing and automobile styles do, soon changed.

The structure that established the new design aesthetic was the George Washington Bridge, which opened in 1931 as the first to span the Hudson River between New York and New Jersey. Although a lower deck would be added in the early 1960s, the original structure had only a single deck that was very shallow and unencumbered by a stiffening truss. Presentation drawings for the bridge, made when it was still a design concept, variously showed both wire-cable and eyebar suspension systems, with the choice expected to be made on the basis of economics. However built, the bridge was expected to become a vital traffic link in the developing road network that would reach from upper Manhattan across to northern New Jersey and on into that part of New York State that is located on the west side of the Hudson. The foundations, towers, cables, and anchorages of the bridge were thus designed eventually to carry two decks, but for reasons of cost at first only one was to be built. Because the deck was made wide, also in anticipation of traffic volume, it was necessarily heavy; engineers were convinced that this feature, in combination with the structure's four massive suspension cables, would provide sufficient stiffness to the roadway to make a truss unnecessary. The resulting design had a 10-foot-deep, 3,500-foot-long main span, which made for a very slender structure indeed. Virtually all suspension bridges designed subsequently in the 1930s strove for this slender, unencumbered look, and by the end of the decade the aesthetic imperative had produced bridges that were so flexible that their decks undulated in the wind and one, the Tacoma Narrows, collapsed in 1940. After that incident, engineers would forgo suspension-bridge building while they rethought safety issues. The 1950s saw a return to more conservative structures and a more pragmatic aesthetic.

In the meantime, Silver Bridge and its clunky contemporaries had carried traffic without incident, even though that traffic had changed in volume and kind. Silver Bridge was designed at a time when the typical automobile was a Ford Model T, which weighed about 1,500 pounds, and the heaviest trucks allowed on West Virginia roads had a gross weight of 20,000 pounds. However, in 1967, when the bridge fell, average family cars weighed around 4,000 pounds, and trucks with a gross weight in excess of 60,000 pounds were using the highways. The bridge's designers never anticipated this effective tripling of the so-called live loads on the bridge. Fortunately, the factor of safety incorporated into Silver Bridge enabled it to support the heavier loads, albeit with a reduced margin. If, in fact, its designers had been asked if the bridge could take the increased loads, they might have expressed concern, but engineers are not always consulted seriously about such commercial matters as vehicle weight or such political decisions as highway weight limits.[7]

West Virginia had conducted a "complete inspection" of Silver Bridge in 1951, but subsequent inspections were of "varying thoroughness," concentrating on repairs made to the bridge deck, sidewalks, and piers, whose concrete was deteriorating somewhat. The eyebars were also looked at, but "from the roadway with the assistance of binoculars," perhaps because the chains were not easily accessible from the bridge deck and perhaps because there was little reason to suspect their integrity. Evidently no more careful inspection of the bridge took place for the next sixteen years. The shocking collapse of Silver Bridge in 1967 prompted Lyndon B. Johnson to establish a President's Task Force on Bridge Safety, headed by the secretary of transportation. The group had three distinct main tasks: one was to determine what caused the failure of the bridge; a second was to plan for a replacement for the fallen structure; and a third was to look into the safety of the nation's bridges generally. This last activity would lead to a rethinking of policy, and in 1968 Congress required the secretary of transportation to establish National Bridge Inspection Standards. In 1970, these were instituted for bridges on highways receiving federal aid, and in 1978 for all bridges of greater than twenty-foot span on public roads of all kinds. Subsequent legislation passed in 1987, following the failure of bridges across the Mianus River in Connecticut and Schoharie Creek in New York State, instituted inspection procedures relating specifically to "fracture-critical" and underwater elements of bridges, those parts that had led to the most recent highly visible failures. Nowadays, every American highway bridge must be inspected at least every two years, and this is one of the positive legacies of the failure of Silver Bridge.[8]

In the immediate aftermath of the Silver Bridge collapse, the first order of business had naturally been to search for survivors of the accident and then to recover the bodies of the victims. In the process of doing this "bridge members were cut apart and thrown about with no regard for anything but getting them out of the way." Only after rescue and recovery was completed could forensic engineers begin the task of identifying what caused the failure. To organize this task, the NTSB established three distinct groups: a Witness Group, whose responsibility was to collect testimony from survivors and eyewitnesses and to interpret it; a Bridge Design and History Group, which was responsible for checking how the bridge was designed, modified, and loaded throughout its lifetime; and a Structural Analysis and Tests Group, whose responsibilities included investigating the bridge's remains and conducting any necessary laboratory tests. These groups were to work together toward a common end in a manner not unlike the way the components of a bridge are supposed to work together to form a structural system.[9]

Getting at the cause of the collapse of Silver Bridge required overcoming many obstacles, not least of which was the fact that the bulk of the superstructure lay in the river, some of it blocking shipping lanes. In the wake of any structural disaster, it is important to record the configuration of the wreckage and to retrieve and preserve as much as possible the various parts without inflicting further damage on them. This is clearly easier said than done when heavy pieces of steel are involved. Because it was imperative to reopen the river to boat and barge traffic as soon as possible, in the case of Silver Bridge "the wreckage was collected and dumped" on twenty-seven acres of field beside the river, where it would lie until it could be organized and scrutinized. After the initial shock of the accident, investigators wisely began to make a photographic record of each piece being lifted out of the water, and this would help them later in understanding what had landed on top of what, invaluable information in determining the sequence of collapse. Other helpful means of placing pieces of wreckage in the proper place were clues provided by such things as paint drippings, which naturally indicated the direction that gravity had acted on the parts when they were in place on the bridge and so could be used to orient a chain link properly in spite of its multiple geometric symmetries. Still, it would take almost as long to reassemble the recovered bridge parts as it did to construct the whole bridge in the first place.[10]

As important as the physical evidence of a failure can be, eyewitness accounts of people at or around the scene of the accident when it occurred can also provide invaluable insights into how the structure behaved just before and during its collapse. Unfortunately, what people remember seeing and how they relate their experience can vary widely. Interviewing witnesses and trying to piece together a consistent narrative of how the failure began and how it progressed can thus be a difficult task. Weeks after what was described as "the most tragic highway bridge accident in U.S. history," it was reported that still "no two eyewitnesses agree on the exact sequence of events during the collapse." One survivor of the Silver Bridge disaster said that "it was all over in seconds," while a later recounting reported that the "collapse occurred in about 60 sec." Early news reports had thirty-one cars falling into the icy water, with casualties estimated to be but "a few." Other sources said at least seventy-five vehicles went down with the bridge. Discrepancies between real-time news reports are common, but they have a tendency to work their way into subsequent reports and studies, which can make the reconstruction of the details confusing at best.[11]

The reported possibility of a loud noise suggestive of an explosion preceding or accompanying the bridge collapse led an army ordnance group to look closely for confirming clues in the condition of vehicles that were removed from the river. Retrieved pieces of the structure were also inspected for damage indicative of the use of explosives to sever them. None was found. Other witnesses believed they had heard a sonic boom. Since the vibrations accompanying such events were known to shatter windows, the possibility that the bridge—which had a reputation for being a shaky structure—had somehow been set into destructive vibration by a boom had to be considered. This, too, was eventually dismissed as not being a credible cause of the collapse.[12]

One early and vague theory was that "something drastic must have happened to the structure the night prior to the collapse, based on reports that the bridge's tremendous population of pigeons had abandoned their home at that time." This notion may have been given some credence because of the superstition that large numbers of birds congregating on a bridge under construction meant that the structure would be safe and could withstand any test to which it might be subjected. If that were true, then it would seem to follow that birds deserting a bridge in droves might suggest that the structure had lost or was about to lose some of its strength, perhaps to the point of collapse.[13]

The investigators of the Silver Bridge failure did not give the bird theory much credence, but half a century earlier it had been the topic of much discussion. In 1907, the Queensboro Bridge was being built in New York City when the Quebec Bridge, then also under construction, collapsed. Since both were cantilever structures, the safety of the Queensboro became the subject of considerable scrutiny. However, the large flocks of pigeons, swallows, and ducks that were roosting on the massive structure each night were offered as evidence of its strength. A Queensboro engineer, Edward E. Sinclair, recalled that in his twenty years of experience building bridges he had never seen so many birds on any other structure on which he had worked. Sinclair quoted ornithologists as saying that "birds in large flocks will not settle on a weak structure," and he also cited the Rudyard Kipling story "The Bridge-Builders" as an authority on the belief that birds on a bridge under construction were an indication that it would pass all tests. (I have not been able to find any such reference in Kipling's story.) At the time, fellow engineers criticized Sinclair for being superstitious, but he defended his position in a letter to the editor of the *New York Times*, stating that newspapers in Quebec had noted the absence of birds from that city's ill-fated structure. Sinclair believed that animals had "some intuitive faculty" regarding the safety of structures. He did admit, though, that he would have maintained a "discreet silence" about the subject had he known it would engender the "very acrimonious debate" that it did at the Engineers' Club.[14]

If the Silver Bridge failure could not be associated with the flight of birds, an explanation might be found elsewhere. One such place was the legend of the Curse of Chief Cornstalk, which had been part of local lore since the American Revolution. During that conflict, Cornstalk was the head of a confederacy of Indian tribes, and in 1774 he led a war party in an attack against a troop of Virginians. Partway through what came to be known as the Battle of Point Pleasant, it became clear to Cornstalk that his forces could not prevail, and so he asked his men whether they wished to fight to the death or surrender. They chose the latter, but Cornstalk never forgot the defeat. Three years after the battle, as he lay dying, he supposedly put a death curse on Point Pleasant and its environs. Whatever bad things happened subsequently, including the collapse of Silver Bridge, were often blamed on the Cornstalk curse.[15]

The heart of a rational failure investigation and analysis consists of framing credible hypotheses that will lead to an understanding of exactly what happened. As with any dramatic structural collapse, even before the last piece of wreckage had been removed there was plenty of early speculation about what specific initiating structural failure brought Silver Bridge down. A professor from Carnegie Mellon University offered that the bridge was overloaded with traffic, and that was why it fell. It was certainly true that the accident occurred during rush hour, but the bridge had been equally or even more crowded on many earlier occasions. Also, as was the case with highway bridges generally, because trucks using Silver Bridge had increased significantly in weight from the time it was designed, the structure had been subjected to increasingly heavy loads. But since the bridge had been loaded to capacity many times before the accident, there had to be something distinctive about conditions on the evening of the failure. What that was, was not immediately apparent.[16]

An article in *Popular Science* magazine blamed "our worst highway disaster" on the design's "radical features," especially its use of eyebar chains. According to an engineer quoted in the article, "if one chain link goes, it all goes." It does not take an engineer to understand this, of course, but in fact all structures have a metaphorical weakest link, and it is the object of design to make sure that link is still strong enough not to break under expected loads and operating conditions. The challenge for the task force investigating the collapse of Silver Bridge was to identify the real or metaphorical weak link that broke and caused the entire bridge to fall into the river—and to explain why that particular link broke when it did.[17]

There were a variety of telling clues. Since the entire bridge had collapsed, the sequence of events must have been precipitated by the failure of, or in, a principal structural component, such as a foundation, pier, tower, anchorage, or suspension chain. The piers, which remained standing, were found to be in sound condition, as reported by divers who also found no scouring around the foundations or any serious damage to them. The piers had also remained in alignment, which ruled out any shifting of the superstructure. The anchorages survived pretty much intact, but the towers had "crumpled" in the collapse. The towers and suspension chains were thus left as the main focus of detailed investigation. But which had failed first? Where did the initiating event occur? In such cases an answer was not always easily arrived at, according to Edward J. Donnelley, a partner in the Greiner design firm, whose wire-cable version of the bridge had not been built, for "it is difficult to separate the primary cause of failure from secondary causes."[18]

A significant configuration of the wreckage pointed to the smoking gun. The downstream eyebar chain fell to the upstream side of the bridge and landed on top of the chain from that side of the structure. This strongly suggested that the chain on the bottom was the one that broke first, throwing the entire bridge out of balance, allowing that side of the roadway to drop, and then pulling the other side down on top of it. This scenario was consistent with several features of the bridge deck and its use. Originally, the bridge was to have three lanes of traffic, but the downstream lane had been converted to a sidewalk, which meant that under normal traffic conditions the upstream chain had to bear a heavier load than its counterpart on the sidewalk side. Thus, assuming both chains were equally strong when installed, it would be the chain carrying the heavier load that could be expected to break first,

SEARCHING FOR A CAUSE   165

possibly due to fatigue. Furthermore, among the fallen chains only a single eyebar link was not in its proper place, and that proved literally to be the weak link.[19]

Less than ten months after the accident, the Task Force on Bridge Safety released an interim report, in which it identified the initiating event of the collapse as a fracture across the eye of the anomalous eyebar associated with the upstream chain. Other members of the bridge structure had also been found to be broken, and the report noted that although "the majority of the fractures don't demonstrate the classic markings of fatigue fractures, the possibility will be explored in the laboratory." The laboratory tests were expected to take at least another nine months. This was understandable, since fatigue tests naturally take time, and to simulate conditions of heavy trucks traversing the bridge throughout its forty-year lifetime as many as ten million cycles of loading and unloading might be required on test specimens. Furthermore, the possibility of corrosion's playing a role in the failure had yet to be addressed. As the forensic engineer Abba Lichtenstein noted in his recounting of the Silver Bridge collapse on the occasion of its twenty-fifth anniversary, "usually, several things have to go wrong at the same time to cause a collapse."[20]

In the 1920s, when Silver Bridge was designed, engineers still had a limited knowledge of the nature of metal fatigue, and the bridge-building industry paid little attention to the possibility of brittle fracture, which occurs with little if any warning. The phenomenon is familiar to anyone who has observed a hardware-store clerk cut a sheet of window glass to size by scoring one side of the sheet and then just bending it away from the scratch. The process happens very quickly and without the kind of precursor movement and noise of fibers snapping that accompany the breaking of, say, a partially sawn-through piece of wood. Recall that brittle fracture had been responsible for the collapse of nineteenth-century railroad bridges like the Dee, the spontaneous rupture in

1919 of the storage tank that flooded Boston with molasses, the break-up of welded Liberty ships during World War II, and the mid-air explosion of Comet jetliners in the mid-1950s. But it was the 1967 collapse of Silver Bridge that gave steel-bridge designers a renewed urgency to learn more about this insidious threat to the integrity of their otherwise graceful and faithful structures.[21]

On close inspection of the brittle-fracture surface on the suspect eyebar from Silver Bridge, John Bennett, of the National Bureau of Standards, noticed that it had two distinct parts. A 1/8-by-1/4-inch-long section of it was covered with rust that was deeply encrusted; the remainder of the fracture surface had only a light coating of rust, which presumably came from its relatively brief immersion in the river. Bennett's belief was that the smaller, more rusted section represented a crack that had grown over a long period of time from some much smaller manufacturing imperfection. The mechanism by which it grew from an imperfection to a flaw was a combination of repeated concentrated forces and corrosion that assisted in the resultant crack's extension into the metal. The processes involved have come to be known as stress-corrosion and corrosion-fatigue crack growth, the first being corrosion accelerated by stress and the second crack growth accelerated by corrosion. When the crack in the eyebar reached a certain critical, albeit still small, size and the load on the bridge chain exceeded the strength of the weakened link, spontaneous brittle fracture occurred on the side of the eye containing the crack. This shifted the entire load that the eyebar had carried to the other (unbroken) side of the eye and caused it to tear apart in a non-brittle way. With the eye broken open, the pin that connected the broken chain link to its neighbor became misaligned and allowed the other eyebar in the link to fall off, severing the chain. With that, the overall structural collapse progressed. The laboratory tests carried out for the Task Force on Bridge Safety confirmed the reasonableness of this hypothesis.[22]

The task force issued its final report a little more than two years after the interim report—and three years and a day after the collapse. It held no real surprises. Rather, it included a methodical recitation of the by-then familiar facts of the case, a description of the analysis carried out to ascertain the cause of the failure, and some conclusions and recommendations. The report contained lots of supporting diagrams, photos, and graphs, some keyed to specific chain links and truss joints by a systematic scheme of identifying letters and numbers. The analysis section included, among other things, a report of the chemical and mechanical analysis of eyebar steel; a comparison of the actual loads on the bridge with those assumed; and a description of "the mechanism of collapse as determined by the process of elimination." The bottom line of the report was contained in a conspicuously short section headed simply "Cause," which stated: "The Safety Board finds that the cause of the bridge collapse was the cleavage fracture in the lower limb of the eye of eyebar 330 at joint C13N of the north eyebar suspension chain in the Ohio side span. The fracture of the structure was the result of the joint action of stress corrosion and corrosion fatigue." Without further elaboration, the report listed three "contributing causes":

1. In 1927, when the bridge was designed, the phenomena of stress corrosion and corrosion fatigue were not known to occur in the classes of bridge material used under conditions of exposure normally encountered in rural areas.
2. The location of the flaw was inaccessible to visual inspection.
3. The flaw could not have been detected by any inspection method known in the state of the art today [1970] without disassembly of the eyebar joint.[23]

It is telling what the safety board did *not* include as a cause (or, as later NTSB terminology would have it, a "probable cause"), or even a contributing cause: There was no mention of negligence on the part of the designers of the structure. They were exonerated implicitly by the observation that, at the time of the bridge's design, stress corrosion and corrosion fatigue were not expected in the kind of steel used in the eyebars. Furthermore, those responsible for inspection and maintenance of the bridge were also absolved of responsibility because the fatal flaw was effectively undetectable. In retrospect, the use of an uninspectable connection detail between eyebars was obviously not a good idea, nor was the use of only two eyebars per chain link. But not being a good design is not exactly synonymous with being a bad design. Had a more corrosion- and crack-resistant material been used for the eyebars, Silver Bridge might be standing today. It was the after-the-fact realization that the eyebar steel was susceptible to corrosion and fatigue cracking, plus the fact that the chain-link joints could not be inspected, that understandably led to the closure and then to the dismantling of the structure's twin, St. Marys Bridge.

The report concluded with recommendations, which closely addressed the limitations that contributed to the cause of the failure. In particular, the secretary of transportation was advised to expand or initiate research programs relating to materials susceptible to slow flaw growth; critical flaw size; inspection equipment; analytical procedures to identify critical flaw locations in structures; standards development; techniques to repair flawed bridges; and gaining further knowledge about the loading and life expectancy of bridges. The report also called for mandatory safety inspections of and federal aid to repair all the nation's bridges, not just those that fell under federal highway aid programs. This is the most significant legacy of what was called the "country's worst bridge disaster."[24]

In retrospect, getting to the cause of the Silver Bridge failure was relatively easy, in that the accident occurred over a river whose flow could be controlled to a degree by an upstream dam and lock,

the closing of which aided not only in the search for bodies but
also later in the recovery of the bridge parts that held evidence relating to the cause of the failure. Accidents over deep-water lakes
or oceans do not always allow such access to clues. In the 1950s, a
series of mid-air explosions of de Havilland Comet jetliners occurred in mid-air over open water, spreading debris over a wide
swath of the deep blue sea. Theories of what was causing the failures ranged from lightning strikes to pilot error. Only the de-
termination and persuasiveness of an engineer who suspected
fatigue-crack growth enabled him to secure a complete Comet
to test in the laboratory. He sealed the fuselage against leaks and
alternately pressurized and depressurized the water that filled it,
thus simulating flight cycles, complete with flapping wings moved
by large hydraulic devices. His theory was that as the planes executed cycle after cycle of loading and unloading, tiny cracks developed and grew—in a manner not unlike the way they would in the
Silver Bridge eyebar—until they were of a critical size that could
not withstand the stresses induced by pressurization. When the
critical conditions of crack size and cabin pressure were reached in
actual flight, the cracks grew so spontaneously and catastrophically that the entire structure broke up into many pieces, making
the sequence of events difficult to reconstruct, even if the pieces
could have been retrieved. That is why, in order to have a more
controlled fracture in the laboratory, the test engineer filled the
fuselage with water rather than with air. He reasoned that when a
crack opened up in the test plane, the water would leak through it
and thus relieve the pressure inside, thereby arresting the breakup
and allowing for a study of the fracture. Had compressed air filled
the fuselage of the test plane, as it does in a plane in actual flight,
the pressure could not have been relieved so quickly, and the crack
and subsequent branching cracks would continue to grow in many
directions. The result could even be an explosive disintegration of
the entire structure, not unlike what happens to a balloon pricked

by a pin. The ensuing flying fragments of the plane not only would have been destructive to the laboratory and its occupants but also would have been damaged in the process and so more difficult to analyze to reconstruct the progress of the failure. The lab tests confirmed the engineer's hypothesis, and the initiating cracks were traced to the squarish corners of the Comet's windows and hatches, which acted not unlike the cavetto corners on the Dee Bridge girders to concentrate stress. Testing hypotheses about accidents is what leads to improved understanding of failure and its causes. The understanding gained from the full-scale Comet fatigue tests has been invaluable in designing fuselages that tear open partially, as they did during the 1988 Aloha and 2011 Southwest airlines flights discussed earlier, but do not explode catastrophically.

Another kind of mid-air explosion occurred on July 17, 1996, off the southern shore of Long Island. It happened only twelve minutes after TWA Flight 800 had taken off from New York's JFK Airport, headed for Paris. Witnesses described seeing the Boeing 747 jumbo jet burst into a fireball, with plane parts raining into the Atlantic Ocean. All 230 people on board were killed. Some witnesses described seeing streaks of light that suggested a missile heading toward the aircraft just before it exploded, and the persistence of the idea that the failure was the result of a terrorist attack eventually led to tests involving rockets being shot into old airplane fuselages sitting in the desert. The resulting damage to the test planes provided guidance on what might be looked for in the wreckage recovered from beneath the water, which was about 130 feet deep around the accident scene.[25]

It took a while, but eventually 98 percent of the wreckage was recovered and patched together to resemble the intact plane. There were no indications that it had been struck by a missile. The NTSB report—issued in August 2000, four years after the accident—declared that the most likely cause was that fumes in one of the plane's fuel tanks were ignited by a spark from a short circuit in some wiring, triggering the fireball that destroyed the aircraft. This conclusion came after numerous tests on mockups of fuel tanks empty except for residual fumes. The conclusion did not satisfy conspiracy theorists, however, and even ten years after the accident they clung to the idea that foul play had brought down the aircraft. One consultant on aviation crashes offered this explanation for such persistence: "It's more acceptable to the public if somebody did this, rather than blame it on maintenance or design."[26]

It is not just theories about the causes of bridge collapses and airplane explosions that evolve over time. White Star, the steamship company that owned the *Titanic*, described the luxurious ocean liner in a brochure as "designed to be unsinkable." Of course, the ship sank on its maiden voyage in April 1912, with the loss of more than 1,500 lives. The exact cause of the *Titanic*'s sinking remained the subject of investigation and debate even after the wreckage was discovered in 1985, two miles beneath the surface of the sea. An expedition a decade later discovered that the starboard bow did not contain the large gash that had long been thought to have been inflicted during the collision with an iceberg. Rather, water had evidently been let into the hull through a half-dozen narrow slits opened up between the ship's steel plates. The nature of the damage led to the theory that the rivets that had held the overlapping plates together had somehow popped during the collision, leading to the flooding of forward compartments, which caused the bow of the ship to tilt down into the water, which in turn allowed more water to flow over the tops of the bulkheads and progressively flood more and more compartments—until the ship could no longer stay afloat.[27]

A combination of metallurgical testing of rivets recovered from the wreck, computer simulations, and archival research in the records of the Belfast shipbuilder Harland and Wolff revealed that inferior rivets may have played a very significant role in the scenario of the slit-open hull. The company's records showed that while it was building simultaneously the *Titanic* and its sister ships *Olympic* and *Britannic,* it was having ongoing problems securing a sufficient supply of rivets and riveters. Each ship required about three million rivets for its completion, and the shipbuilder had to obtain some from smaller forges, whose quality control might not have been as stringent as that of its larger and regular suppliers. A further problem was that in some cases No. 3 iron bar, designated "best," was ordered instead of the customary No. 4, or "best-best," for use in making rivets. Also, the ships were being built at a time when the use of iron rivets was giving way to steel ones, and in fact the *Titanic* contained a mixture, iron rivets being used mainly in the bow and stern sections and steel ones amidships, where stresses on the structure were expected to be more intense. Indeed, the damage to the ship's hull, investigators discovered, "ends close to where the rivets transition from iron to steel."[28]

Subjecting the iron rivets recovered from the *Titanic*'s wreckage to modern testing techniques, metallurgists found that they contained large amounts of slag, which rendered them brittle and more easily fractured, especially in the cold ambient temperatures of the North Atlantic. All of this may not have been known in theoretical terms in the early twentieth century, but by experience shipbuilders then had learned that "best-best" iron was the kind to use in critical ship parts and equipment, such as anchors, chains, and rivets. Compromising on the quality of iron, for whatever reason, may indeed have built into the *Titanic* vulnerabilities that it could not afford.[29]

As infused with drama and tragedy as the story of the *Titanic* is, the story of another ship, the *Edmund Fitzgerald,* has become legendary around the American Great Lakes. This investment venture of the insurance firm Northwest Mutual was named for the chairman of the board of the Northwest Mutual Life Insurance

Company. The *Edmund Fitzgerald* was the largest ship on the lakes when launched in 1958 and was expected to remain among the largest because of the limitations imposed by the St. Lawrence Seaway locks. Its primary cargo was expected to be taconite, a low-grade iron ore found in Minnesota and shipped by lake freighters to the steelmaking centers around Lake Erie.[30]

On November 9, 1975, the *Edmund Fitzgerald* left Superior, Wisconsin, carrying twenty-nine crewmen and about 26,000 tons of ore. A storm from the plains had been proceeding toward the lakes, and a gale warning had been issued for Lake Superior. This caused the ship's captain to sail closer to the Canadian shore than he might otherwise have done, to take advantage of the sheltering effect of the land and thereby to escape the larger waves expected farther out. With shifting winds, the ship developed a list. This much is known from the *Edmund Fitzgerald* captain's communications with the captain of another ship on the lake, who watched the fated ship suddenly disappear from his radar. In the absence of eyewitnesses, the investigators preparing a marine casualty report had to reconstruct a probable cause of the accident based on the configuration of the wreckage. According to the report, as the ship encountered larger waves, they washed over the deck and dumped water into an incompletely sealed hold. As more and more water filtered through the cargo of ore pellets, the ship rode lower in the water, allowing even more waves to wash over the deck. Eventually, one large wave was hit head on, causing a jolt to the ship. When this happened, the cargo in the hold surged forward, pulling the bow down into the water, where it dived quickly and hit the bottom with such force that the midsection of the ship crumpled and the stern section flipped over the bow and landed upside down in front of it. The Great Lakes Shipwreck Historical Society has sent three expeditions to the bottom of the lake, but though divers retrieved the ship's bronze bell in 1995, no further information on the cause of its failure has yet surfaced. According to the Great

Lakes Shipwreck Museum, the sinking of the *Edmund Fitzgerald* "remains the most mysterious and controversial of all shipwreck tales heard around the Great Lakes," and "her story is surpassed in books, film and media only by that of the Titanic."[31]

Some failures do become legendary, even if their cause remains somewhat uncertain. But it is the failure whose cause is so well established as to be virtually incontrovertible that should be the most widely studied—not just because it was a failure but because it holds a clear lesson. The story of Silver Bridge and its failure is in this category. The outstanding attention to detail exhibited in the accident investigation, especially with regard to the hard evidence of the fractured eyebar, left little doubt that the failure was rooted in a design that inadvertently made inspection all but impossible and failure all but inevitable. If ever a design was to blame for a failure, this was it. The intention of the bridge's designers was evidently nothing if not honorable. They did not wish to design a doomed bridge; they did so because they—and all their contemporaries—were ignorant of the full implications of the materials and details they employed. They were also perhaps overly cavalier about relying so much on eyebar links that were cleverly placed to do double duty. In retrospect, the designers should not have proceeded in such ignorance, and they should have realized the potentially catastrophic consequences of their design choices. But this was a different time. If there is anything positive about the Silver Bridge failure, it is that its legacy should be to remind engineers to proceed always with the utmost caution, ever mindful of the possible existence of unknown unknowns and the potential consequences of even the smallest design decisions. The story of Silver Bridge is a cautionary tale for engineers of every kind.

EIGHT

The Obligation of an Engineer

Engineers are always striving for success, but failure is seldom far from their minds. In the case of Canadian engineers, this focus on potentially catastrophic flaws in a design is rooted in a failure that occurred over a century ago. In 1907 a bridge of enormous proportions collapsed while still under construction in Quebec. Planners expected that when completed, the 1,800-foot main span of the cantilever bridge would set a world record for long-span bridges of all types, many of which had come to be realized at a great price. According to one superstition, a bridge would claim one life for every million dollars spent on it. In fact, by the time the Quebec Bridge would finally be completed, in 1917, almost ninety construction workers would have been killed in the course of building the $25 million structure.

The 1907 collapse alone claimed the lives of seventy-five workers. That accident occurred because the design engineer did not properly anticipate or correctly calculate the weight of the structure; because the resident site engineer did not heed warnings that the steel was overstressed; and because the chief consulting engineer *in absentia* inappropriately delegated authority to inexperienced younger engineers. By keeping the causes and lessons of the Quebec Bridge failure very much in mind and by keeping a re-

The wreckage of the Quebec Bridge, which collapsed in August 1907 while still under construction, littered the banks of the St. Lawrence River. The accident, which claimed the lives of seventy-five construction workers, became symbolic of the dire consequences of careless and inattentive engineering. The Canadian iron-ring tradition, which dates from the early 1920s, is symbolically linked to this failure.

minder of it literally close at hand, Canadian engineers today are less likely to make errors of their own in the design office or in the field. These days, increasing numbers of American engineers have been emulating the tradition of the Canadians, but unfortunately the Americans have not been so explicitly linking their efforts to failure, either to a specific incident or to a general principle, either in theory or in practice, either in or out of school.[1]

Like many long-established engineering programs, Yale's "scientific school," which evolved into the university's School of Engineering, dated from the mid-nineteenth century. Among Yale's early distinctions is having awarded, in 1863, the first engineering doctorate granted in America. It was conferred upon Yale's first engineering doctoral student, Josiah Willard Gibbs, who went on to become one the country's most distinguished scientists. Yet the title of his thesis, "On the Form of the Teeth of Wheels in Spur Gearing," may have given little hint that he would become a professor of mathematical physics at Yale and, through his applications of thermodynamics to chemical reactions, lay the basic theoretical foundations for the discipline of physical chemistry. Gibbs contributed much to the development of the field of statistical mechanics, which unites conceptually mechanical and thermal phenomena at the molecular and macroscopic scales, thereby relating knowledge about atoms and molecules to the bulk properties of materials. In the early twentieth century, Gibbs proposed the more specific name "statistical thermodynamics" for the relatively new field.[2]

But the distinguished roots of Yale's engineering school did not insulate it from insult; its early successes did not count against later failures, at least in the minds of some administrators. In 1994, the Yale alumni magazine published an article on the rebuilding of engineering at the Ivy League university. The development was newsworthy because it represented an about-face for the institution. In 1963, during fiscally difficult times, President Kingman Brewster, Jr., had replaced engineering's traditional divisional structure at Yale by consolidating its departments under the rubric "engineering and applied science." Two decades later, President Benno Schmidt had targeted engineering, among other programs, for drastic restructuring and major faculty reductions. Some observers feared that engineering at Yale was on its way to extinction. As it turned out, Schmidt himself became a casualty of his own plan, which led to the naming of a new president, Richard Levin, who in turn announced an institutional commitment to rebuilding Yale's leadership in engineering education and research. Levin also made a clear statement about the broader importance of his decision: "As technological change shapes the world in which we live, a university that aims to educate leaders for our nation and for the world must nourish the study of engineering and applied science." It is interesting to speculate on the circumstances that may have contributed to the quick about-face on engineering at Yale.[3]

When Schmidt announced Yale's cutback in engineering, the university's distinguished physics professor and director of its nuclear structure laboratory, D. Allan Bromley, was on extended leave to serve the government in Washington, D.C., having been appointed chief science and technology advisor in the George H. W. Bush administration. In that capacity, Bromley naturally crossed paths with many influential engineers, some of whom may have been Yale alumni but many of whom must have suspected as soon as they met Bromley that he had to be a kindred spirit, even if they knew nothing of his background or politics. The small detail that marked him as an engineer—and most likely a Canadian one at that—was evident in the striking full-page, full-color portrait of him that accompanied the alumni magazine article.[4]

But what is it in Bromley's appearance that the informed observer saw? In his portrait, Bromley is dressed fashionably in what looks to be a navy blazer over a dark-striped shirt with white cuffs and collar, closed off by a red bow tie—colors that the photographer no doubt saw echoed dramatically in the large abstract painting before which Bromley stands. In front of the bold background, the dapper Bromley appears to be the image of control and self-assurance, his somewhat askew bow tie and slightly misfolded breast-pocket handkerchief just barely offsetting the serious expression on his face. His left hand is all but hidden in his pants pocket, but his other hand, projecting out from just the right amount of shirt cuff to balance the white of the handkerchief, holds his metal-rimmed glasses, sharply defined against his dark

D. Allan Bromley (1926–2005), dean of engineering at Yale University when this portrait was taken, proudly wore a pinkie ring that marked him as an engineer, and one most likely educated in Canada. The common belief that the faceted iron ring he received upon participating in the Ritual of the Calling of an Engineer was made from the wreckage of the Quebec Bridge is belied by the fact that that ill-fated structure was made not of iron but of steel. In time, most rusting iron rings were replaced by stainless steel ones. Since the 1970s, the American Order of the Engineer has emulated the Canadian Iron Ring tradition, but its stainless steel rings have a more gently curving profile.

jacket. All this could describe any confident leader standing for a portrait, but in holding his glasses thus, Bromley turns the back of his hand outward and displays on his little finger the one thing that marked him as an engineer: a stainless-steel pinkie ring. It is the presence of this ring that very well might have prompted other engineers in Washington to see Bromley as more than just another well-dressed political appointee from the science side of the science and technology divide. This unpretentious ring would have signaled to knowledgeable engineering leaders that Bromley would fully appreciate their assertions that Benno Schmidt's and Yale's treatment of engineering was "insulting to the profession" and that something had to be done.[5]

Remarkably, Bromley had established his considerable reputation on the basis of his work not as an engineer but as a scientist. He held a doctorate in nuclear physics from the University of Rochester and had been associated with teaching and research in that field at Yale for decades before going to Washington. But, just as J. Willard Gibbs came to science from engineering, so did Bromley—and he never seemed to have forgotten it. In fact, the ring sported by Bromley is explained by his Canadian engineering roots. He was born in 1926 in the tiny village of Westmeath, in northeastern Ontario. He attended Queen's University, in Kingston, Ontario, and in 1948 received his bachelor's degree in engineering physics from that school's Faculty of Engineering. In addition to receiving his diploma, he participated with his classmates in a private, little-publicized event known colloquially as an Iron Ring Ceremony but officially designated as the "Ritual of the Calling of an Engineer." At the ceremony, Bromley would have recited the "Obligation of the Calling of an Engineer," a declaration of professional standards and honor, signifying a rite of passage not unlike that undertaken by medical doctors upon their graduation from medical school.[6]

Although almost universally known as the Hippocratic Oath, the medical profession's twenty-five-centuries-old statement of commitment to ethical practice is believed by scholars to have been more closely associated with followers of Pythagoras than with those of Hippocrates. Regardless of its attribution, the text has been adapted over the centuries for different times and cultures, but certain dated references and timeless principles survive even in a twenty-first-century translation. Thus, until relatively recently, modern doctors have sworn by "Apollo the physician" to uphold the oath and to, among other things, "do no harm" and "not use the knife" unless they are "trained in this craft." Engineering codes of ethics, which generally date from the early twentieth century, echo much in the doctors' oath. In particular, the fundamental canons of the Code of Ethics of the American Society of Civil Engineers call on its members to "hold paramount the safety, health, and welfare of the public" and to "perform services only in areas of their competence."[7]

Such principles are implicit in the commitments that Canadian engineers make to their profession and are symbolized in the Iron Ring. The ring that a young Allan Bromley received at the Canadian engineers' ceremony was made of wrought iron, and over time it rusted so badly that he—like others of his generation—replaced it with the stainless-steel version that he wore in his Yale portrait. (Even symbols of failure can fail!) The photo of a mature Bromley shows his ring fitting tightly on his right pinkie finger, suggesting that he did not remove it often, if at all. The ring's presence further suggested that Bromley was right-handed and that he continued to think of himself as an engineer, for by tradition the Iron Ring is worn on the little finger of the working hand, but only as long as the wearer is still considered part of the engineering profession. Although the Iron Ring tradition has been almost universally adopted by Canadian engineers, participating in the ceremony and wearing the ring are not necessary to practice engineering in Canada. However, those who do take part rarely remove the ring until they leave the profession, at which time, again by tradition, the ring is surrendered. Some Iron Rings have been passed on from older family members, or from mentors, to younger engineers as part of the ceremony.[8]

The Iron Ring Ceremony has its origins in the early 1920s, when Herbert E. T. Haultain, a professor of mining engineering at the University of Toronto, wished to improve the image of the profession among engineers themselves. He hoped to institute a ceremony akin to that in which young doctors took the Hippocratic Oath. After graduating from Toronto in 1889, Haultain worked in the mining industry in Europe and British Columbia, where he gained firsthand knowledge of conditions that convinced him of the need to instill high ethical standards in young engineers. The conditions Haultain lamented included that of practicing engineering without the proper expertise, which naturally increased the probability of failure resulting from inadequate design. The untrammeled practice of engineering was widespread in the early part of the twentieth century, as the adoption of codes of ethics by professional societies and the institution of professional registration during the period attest. Haultain also wished to see an organization formed that would unite all Canadian members of the engineering profession, regardless of their specialty.[9]

In 1922, Haultain spoke about his concerns at a meeting in Montreal attended by seven past presidents of the Engineering Institute of Canada, the successor to the Canadian Society of Civil Engineers, which was more inclusive than its name implies. The term *civil engineer* had begun to be used in the late eighteenth century to refer to all engineers who were not associated with the military. In the middle of the nineteenth century, with the development of such technologies as the railroads and the telegraph, engineers in many countries increasingly distinguished themselves with classifications like mechanical, electrical, and mining engineers and, where the numbers allowed, formed their own specialized professional societies. In Canada, where there was a smaller population of engineers than in the United States, for example, the Canadian Society of Civil Engineers continued to count all non-military engineers among its members, but there was increasing opposition to doing so. The society thus changed its name to the Engineering Institute in the interests of inclusivity. At the Montreal meeting, the past presidents embraced Haultain's ideas for further uniting the profession and encouraged him to pursue them.[10]

Haultain wished to establish a personal and formal inducting ritual for young engineers who had not yet fully encountered the real world beyond the classroom, and so he wrote to Rudyard Kipling, who happened to be in Canada at the time, asking him to compose the words for a ceremony. Kipling had long been both the literary champion and hero of engineers, having published such sympathetic pieces as the short story "The Bridge-Builders" in the Christmas 1893 number of the *Illustrated London News* and the poem "The Sons of Martha" in 1907. "The Sons of Martha" is rooted in the Gospel text of Luke (10:38–42), in which Christ, visit-

ing the house of Martha, approved of her sister Mary's sitting and
listening to him teach rather than helping Martha, who was busy
serving everyone. Kipling identified engineers with Martha and
her children, who continued to do the chores necessary to keep a
household running rather than sit at the Lord's feet and listen to
his wisdom, as did Mary, and presumably her sons and daughters.
The opening stanzas of the poem convey its tone:

> The Sons of Mary seldom bother, for they have inherited
>   that good part;
> But the Sons of Martha favour their Mother of the careful
>   soul and the troubled heart.
> And because she lost her temper once, and because she was
>   rude to the Lord her Guest,
> Her Sons must wait upon Mary's Sons, world without end,
>   reprieve, or rest.
> It is their care in all the ages to take the buffet and cushion
>   the shock.
> It is their care that the gear engages; it is their care that the
>   switches lock.
> It is their care that the wheels run truly; it is their care to
>   embark and entrain,
> Tally, transport, and deliver duly the Sons of Mary by
>   land and main.[11]

Although some later engineers would read Kipling's poem as
condemning engineers to second-class status compared with managers, those of Haultain's generation were pleased to take "The
Sons of Martha" as their defining text. Kipling's response to Haultain's invitation was enthusiastic, and he drafted, in consultation
with Canadian engineering groups, the "Ritual of the Calling of an
Engineer," which contains the "Obligation" by which engineers es-
chew poor workmanship and "honourably guard" the reputation

of their profession. Each participant in the ceremony would be able to frame and exhibit a personal, signed copy of the "Obligation," but the content of the Ritual otherwise was to be "neither for the public nor the press." In an attempt to make the prohibition enforceable, the text was copyrighted. (In time, the ceremonies would become open to the parents of participants and so can no longer be considered secret.)[12]

The idea of an inductee's signing an "Obligation" did not originate with Haultain or Kipling. Indeed, the Royal Society, founded in London in 1660, continues to this day a tradition of having each new fellow and foreign member sign its Charter Book. The book was created in 1663, when the Royal Society was granted its second royal charter, which established the society's structure. At the head of each signature page appears "The Obligation of the Fellows of the Royal Society":

> We who have hereunto subscribed, do hereby promise, that we will endeavour to promote the good of the Royal Society of London for improving Natural Knowledge and to pursue the ends for which the same was founded; that we will carry out so far as we are able, those actions requested of us in the name of the Council; and that we will observe the Statutes and Standing Orders of the said Society. Provided that, whensoever any of us shall signify to the President under our hands, that we desire to withdraw from the Society, we shall be free from this Obligation for the future.[13]

In like manner, Kipling's Ritual, as the "Ritual of the Calling of an Engineer" has sometimes been called, was "instituted with the simple end of directing the newly qualified engineer toward a consciousness of the profession and its social significance, and indicating to the more experienced engineers their responsibilities in welcoming and supporting the newer engineers when they are ready to enter the profession," according to the Iron Ring website. The first Iron Ring Ceremony was held in 1925, and the first rings were made of unpolished "hammered iron" that Kipling called "cold." The poet also said of the unpolished, faceted ring that it was rough as a young man's mind, its edges not yet smoothed off. Although some say the writer used the adjective "cold" because the structural material did not forgive the mistakes of engineers working in it, the opening stanza of his poem "Cold Iron" puts it in a different and more positive context:

> Gold is for the mistress—silver for the maid—
> Copper for the craftsman cunning at his trade.
> "Good!" said the Baron, sitting in his hall,
> "But Iron—Cold Iron—is master of them all!"[14]

According to one engineer, the Iron Ring "binds the engineer allegorically to the profession." The ring's circular shape has been said to symbolize the continuity of the profession and its methods, and the circle is also an appropriate symbol of the engineering design process, which is iterative and can seem hopelessly self-referential to the uninitiated. Tradition has it that the rings were fabricated from the wreckage of some catastrophic engineering failure. Bromley believed that his original ring came from the remains of the ill-fated Quebec Bridge, because that is what the dean of the Faculty of Engineering at Queen's University, who gave Bromley his ring, told him. After the accident, the bridge was redesigned, but it suffered a second mishap in 1916, when its center span dropped into the water while being hoisted into place, further embarrassing the engineering community. Finally, in 1917, the bridge was completed, and it continues to stand across the St. Lawrence River—a symbol of Canadian engineering and national resolve. It has also been a symbolic gateway under which immigrants have sailed into Canada. The bridge remains today the longest cantilever span in the world and is a reminder to Canadian engineers, especially those who wear the Iron Ring, to take care with design and construction and to persevere in the face of adversity.[15]

That the Iron Rings were made from the wreckage of the Quebec Bridge is surely apocryphal, for the bridge was made not of wrought iron that could easily be hammered into a faceted band but of harder and stronger steel. In fact, claims and statements about the true source of the first rings are legion. According to one story, the original rings were made by Canadian World War I veterans participating in an occupational-therapy program in a Toronto veterans' hospital, using common iron-pipe stock, and such stock is said to have long remained the standard material. Another source states that some early rings were made from a wrought-iron railroad-track spike from Kamloops, a city in south-central British Columbia. Both may be true, for participants in different parts of the country may have found different sources for the material of their rings. Whatever the origins of the physical ring, however, the Iron Ring is rich in symbolism. Its facets were said to be a "sharp reminder" of an engineer's obligation to careful work, and they were especially effective on a new ring, where they could be "nearly sharp enough to be considered serrations." As the original rings smoothed with age, so it was suggested that with experience will the rough edges disappear from a young engineer's mind, leaving a more integrated wisdom. The ring has also been seen as "a symbol of humility; as the facets of iron wear down over the years, so too the engineer's professionalism mellows with experience." Allan Bromley, who had gained his earliest engineering experience in Canada's hydroelectric-power industry and extended it later at Yale, had clearly earned the right to replace his rusting Iron Ring with the stainless-steel one that had in the meantime become the standard.[16]

Whatever the material of the original rings or of the most recent ones, they are all called "Iron Rings," and their tradition is firmly institutionalized. The "Ritual of the Calling of an Engineer" was endorsed in 1922 by the seven past presidents of the Engineering Institute of Canada, who formed the Corporation of Seven Wardens charged with the administration of the ritual. (The past presidents were named Founding Wardens of the Corporation.) Local groups called Camps were formed, each with seven wardens who were practicing engineers. Not surprisingly, Camp 1 was established at the University of Toronto. Today, four universities from that city's metropolitan area are associated with the Toronto Camp: Ryerson University, the University of Ontario Institute of Technology, the University of Toronto, and York University. This original camp is the only one that continues to offer the option of a wrought-iron ring. Although the camps are associated geographically with college and university cities and towns, they remain independent of any individual academic or other Canadian institution.[17]

In time, the Canadian Iron Ring tradition came to be known among engineers in the United States. I first learned of it in the 1960s, when I was a graduate student and most rings were still made of wrought iron. Like other large Midwest schools, the University of Illinois attracted a lot of Canadians to its graduate engineering programs, and the Department of Theoretical and Applied Mechanics was no exception. As a young graduate student from New York, I was keenly aware of the wide variety of non–East Coast accents I heard around the department, the campus, and the town. These made me acutely conscious of my own accent in that virtually foreign land. The Canadian accents were especially noticeable, but also somewhat familiar, for they vaguely reminded me of the speech characteristics of my upstate relatives in Rochester, which is of course located hundreds of miles inland from the Atlantic coast and on the south shore of Lake Ontario, just a ferry ride across from Toronto. Before too long I noticed that the Canadian engineering students, already distinguished by the sound of their voice, were also distinguished by the oddly faceted iron ring that they wore on their little finger. Virtually unpolishable, it did not look like a piece of jewelry, and I became increasingly curious about its significance. As I got to know my Canadian colleagues better, I asked one of them about the ring and thus learned the rudiments of the Iron Ring tradition. Over beers late at night, almost anything could become an open secret, but I still learned then only the very basics of the Iron Ring tradition.

I certainly was not the first American engineer to become interested in the Iron Ring of Canadian engineers, nor was I the first to learn of its significance. In the early 1950s Lloyd A. Chacey, a civil engineer and executive secretary of the Ohio Society of Professional Engineers, wrote to the Corporation of Seven Wardens inquiring about the possibility of extending the Iron Ring Ceremony below the Canadian border. Although copyright was said to be among the impediments to such a move, it is easy to imagine that the Canadians did not wish to dilute the proud and distinctly indigenous tradition that was uniquely theirs. Correspondence between Chacey and the Corporation continued, however, and in 1962 Homer T. Borton and G. Brooks Earnest, two officers of the Ohio Society of Professional Engineers, were invited to participate in the Canadian ceremony, thereby gaining a model for establishing their own at home. By the mid-1960s, a group of Ohio engineers was pursuing the establishment of an independent Order of the Engineer in the United States.[18]

The late 1960s were especially trying times in America, and engineers were beginning to be attacked as proponents of war and enemies of the environment. The political climate, and perhaps their personalities, encouraged engineers to circle their wagons around the profession, and the disruptive atmosphere on campuses and in political arenas generally made it difficult to forge initiatives rooted in tradition. In 1970, however, an incident of student unrest at Cleveland State University—at the time, campus antiwar protestors around the country tended to view engineers as complicit in weapons proliferation—prompted some engineering student leaders to look for a means of asserting some more positive values. Dean Burl Bush, who had been working with Borton and Earnest in trying to establish a ring ceremony, described the idea to the engineering students. Within three weeks, they had used a metal lathe to turn some rings out of stainless-steel tubing and had organized the first steel-ring ceremony in the United States. It took place on June 4, 1970, and some 170 engineering seniors and faculty members, each of whom made his own ring, participated. Shortly thereafter, the Order of the Engineer was formed as a national organization, with Lloyd Chacey, who had done so much to institute a ring ceremony in the United States, serving as secretary of its Board of Governors.[19]

The express purpose of establishing the Order of the Engineer was "to foster a spirit of pride and responsibility in the profession, to bridge the gap between training and practice, and to present the public with a visible symbol identifying the engineer." With the Cleveland State precedent, other groups of engineers and engineering students began to hold Order of the Engineer ceremonies, with the earliest ones taking place around Ohio. The local chapters were known as Links, and the ceremony was clearly modeled after the Canadian Ritual, but without the benefit of authorship of a Kipling. In 1972, the custodians and administrators of the Canadian Ritual inquired into the workings of the new Order of the Engineer and, after inspection of the manual and ring used, concluded that "they do not infringe on the Corporation's copyrights or patent." The chief warden then conveyed the corporation's pleasure at the reference to the Canadian Ritual in the newer American ceremony. He also conveyed to the members of the Order wishes of "every success in their endeavor to advance the feeling of fraternity" among engineers.[20]

By the mid-1980s, the fledgling steel-ring ceremony had taken place in more than thirty states, with tens of thousands of young and older engineers having recited the American "Obligation." On several occasions I happened to be at a meeting where a ring ceremony took place, and I attended as a very interested observer. The public ring ceremonies all followed more or less the same script, with presentations on the purpose and history of the Order of the Engineer and on the significance of the ring. Those engineers receiving rings then signified their acceptance of the American "Obligation of an Engineer" by reciting it aloud. In its original wording, as posted on the website of the Order of the Engineer, it read as follows:

> I am an Engineer. In my profession I take deep pride. To it I owe solemn obligations.
>
> Since the Stone Age, human progress has been spurred by the engineering genius. Engineers have made usable nature's vast resources of material and energy for Mankind's benefit. Engineers have vitalized and turned to practical use the principles of science and the means of technology. Were it not for this heritage of accumulated experience, my efforts would be feeble.
>
> As an Engineer, I pledge to practice integrity and fair dealing, tolerance and respect, and to uphold devotion to the standards and the dignity of my profession, conscious always that my skill carries with it the obligation to serve humanity by making the best use of Earth's precious wealth.
>
> As an Engineer, in humility and with the need for Divine guidance, I shall participate in none but honest en-

terprises. When needed, my skill and knowledge shall be given without reservation for the public good. In the performance of duty and in fidelity to my profession, I shall give the utmost.[21]

A booklet distributed at the ceremonies during the 1980s stressed God and country and contained color photos of the American flag, fireworks, and the iconic Earth view taken from Apollo 8. The "Obligation of an Engineer" was printed on the booklet's back cover. In time and in deference to heightened sensitivities, the reference to "Mankind's benefit" was changed to "Humanity's benefit," and the phrase "in humility and with the need for Divine guidance" was deleted. Otherwise, the American "Obligation" remains essentially as originally recited.[22]

After their recitation of the "Obligation," the participating engineers were presented with their stainless-steel rings, which, unlike the faceted Canadian rings, are plain and smooth. This part of the ceremony explained the purpose of a larger than one-foot-diameter model of the ring that had been standing upright on a table to the side of the lectern. As I later discovered, dimensioned drawings sufficient for fabricating the wooden ceremonial ring are given on the organization's website, along with details of painting it with "3 coats plastic engineering #960 aluminum epoxy" and mounting it by means of a vertical dowel on a wooden base, which is to be painted "flat black." From a distance, the finished wooden ring does look metallic. But the ceremonial ring is not intended to just sit there as a symbol of the proceedings; it is actually an integral part of them. As each engineer's name is called out, he or she approaches the table and sticks his or her working hand through the large ceremonial ring so that a pre-ordered stainless-steel ring can be placed on the fifth finger of that hand. The potentially confusing logistics of getting the right size ring on the appropriate hand were highlighted at one ceremony when, in order to expedite the proceedings, the relevant ring size was announced with each engineer's name.[23]

The Order of the Engineer, in a manner similar to its Canadian precursor, "is not a membership organization; there are never any meetings to attend or dues to pay. Instead, the Order fosters a unity of purpose and the honoring of one's pledge lifelong." Thus engineers who participate in the ring ceremony are not likely to have throughout their career any further formal association with the Order itself, unless perhaps they participate in some future ceremony as a host or as an observer.[24]

I have never had an opportunity to observe a Canadian Iron Ring Ceremony, and even after many beers I never did ask my Canadian colleagues exactly what took place at one. However, I did come across the wording of the "Obligation of the Calling of an Engineer" in an introductory engineering textbook published in New York in 1994. Years later, I also found the text of the Canadian "Obligation" on the Internet, contained in a PowerPoint presentation prepared to give Class of 2006 graduating engineering students at the University of Alberta a preview of what they should expect in the ceremony in which they were soon to participate. Except for the word changes of "man" to "mankind" and "brothers" to "colleagues," and allowing for each student to insert his or her name where indicated, the two versions agreed verbatim:

> I, [name], in the presence of these my betters and my equals in my Calling, bind myself upon my Honour and Cold Iron, that, to the best of my knowledge and power, I will not henceforward suffer or pass, or be privy to the passing of, Bad Workmanship or Faulty Material in aught that concerns my works before mankind as an Engineer, or in my dealings with my own Soul before my Maker.

My Time I will not refuse; my Thought I will not grudge; my Care I will not deny towards the honour, use, stability and perfection of any works to which I may be called to set my hand.

My Fair Wages for that work I will openly take. My Reputation in my Calling I will honourably guard; but I will in no way go about to compass or wrest judgment or gratification from any one with whom I may deal. And further, I will early and warily strive my uttermost against professional jealousy or the belittling of my working-colleagues, in any field of their labour.

For my assured failures and derelictions, I ask pardon beforehand of my betters and my equals in my Calling here assembled; praying that in the hour of my temptations, weakness and weariness, the memory of this my Obligation and of the company before whom it was entered into, may return to me to aid, comfort and restrain.

Upon Honour and Cold Iron, God helping me, by these things I purpose to abide.[25]

Among the slides in the Alberta digital presentation were two informing the students what to expect on the day of the ceremony: that they would be required to wear business attire; that they would be luncheon guests of the Association of Professional Engineers, Geologists and Geophysicists of Alberta, but that no guests were invited to lunch; that the ceremony would take place immediately after the meal; that the ceremony was open to obligated engineers and to two guests of each candidate engineer; that rings were to be picked up prior to the ceremony; and that no photographs or latecomers would be allowed. Another slide provided some insight into the ceremony itself, which would be presided over by the seven wardens of the camp.[26]

As described in the Alberta slideshow, the ceremony would open with seven strokes on an anvil, representing the driving home of a rivet, as would have been done in constructing the Quebec Bridge. The struck anvil would ring out in Morse code the letters S-S-T, which are said to stand for "Stone, Steel, and Time, which proves them," and also for "Soul, and the Spirit of Man, which Time tries." After some introductory words, there would be a reading from the Book of Esdras, "to impress upon us the limitations of human knowledge, and cultivate humility," followed by the joint recital of the "Obligation" and the placing of rings on fingers by the wardens. This would be followed by a reading of a Kipling poem, either "The Sons of Martha" or the "Hymn of Breaking Strain," which was first published in *The Engineer* and which begins as follows:

> The careful text-books measure
> (Let all who build beware!)
> The load, the shock, the pressure
> Material can bear.
> So, when the buckled girder
> Lets down the grinding span,
> The blame of loss, or murder,
> Is laid upon the man.
> *Not on the stuff—the man!*[27]

This fitting end to a ceremony during whose introduction the Quebec Bridge was no doubt mentioned would certainly bring the minds of the engineers back to the idea of failure and responsibility. With the tone set, the ceremony could be declared finished by the same solemn seven strokes on the anvil with which it had begun.[28]

Rumor has it that participants in the Canadian Ritual usually hold onto a chain during the ceremony, and that the chain is attached to an artifact representing an engineering failure. Perhaps this might be a rivet or other piece of steel from the original Quebec Bridge, or it might be a fractured piece of some other failed structure that had a special symbolic meaning for a particular camp. Whatever the tangible connection might be to failure, the metaphorical connection is no doubt much stronger and evocative of the origins of the Iron Ring Ceremony.[29]

The Canadian and American Obligations, composed almost a half-century apart, reflect in their focus and emphasis and tone the different concerns of different times and different cultures. In the early 1920s in Canada, the national tragedy of the Quebec Bridge failure was still fresh in the minds of citizens and, especially, of engineers, who were embarrassed and humbled by it. World War I, which had introduced the concept of airborne battle and bombing, was also fresh in the minds of people everywhere, but mostly in Europe, and so it should come as no surprise to find the Kipling "Obligation" sounding prayer-like and petitioning. The British, a people of whom Kipling was an exemplar, historically had been dogged by structural failures, especially of railroad bridges like the Tay, remembered to this day throughout Britain for its questionable design, workmanship, material, and maintenance—dishonorable acts attributed to the bridge's engineer and those he supervised. In Canada, as in America at the time, questionable practices by and among engineers were making the case for professional registration, codes of ethics, and honor.

In contrast to the humble tone of the Canadian "Obligation," in which the word *honour* occurs four times, the American "Obligation" is a celebration of engineers and engineering. It begins as a proud statement, boasting of progress and genius and beneficence, perhaps in response to the antitechnology protests of many college students and faculty members who saw engineers and their creations as complicit in warfare and harmful to the planet. Written amid the environmental awakening of its time, it speaks of using the engineer's skill to make "the best use of Earth's precious wealth" and expresses no reservations that this will be done. The concept of failure is explicitly absent from the American "Obligation." World War II had occurred a generation earlier and far from American soil. Engineering and scientific achievements had been instrumental in winning that war. In the postwar era, especially in the academic settings in which obligations are drafted, support for science and engineering would bring windfall benefits to research and development programs. Just a year before the inaugural Order of the Engineer ceremony at Cleveland State, technology would take astronauts to the Moon. The future looked bright for science and engineering, even as there were increasing demands to consider the ethical implications of weapons of mass destruction and environmental insults to the planet.

Of course, the Kipling "Obligation" was written by a poet and writer, and one who had been effectively commissioned by an impassioned engineer to unify the profession and make it honest. The American "Obligation" was written by "a young CSU professor" whose dean asked him "to prepare a creed and a ritual to enhance the dignity and meaning of the ceremony." The professor, John G. Janssen, and his wife, Susan, developed the American "Obligation of an Engineer." The principal inspiration seems to have been the emulation of the Canadian tradition of the Iron Ring and its obligating ceremony. That was a noble objective, but because the copyrighted "Obligation" could not simply be replicated there was necessarily a struggle for distinction. The resulting statement was assertive rather than imploring, and over time the popularity of the American ring ceremony grew with the optimism and failure-free language of its "Obligation." Nevertheless, the most recent ring ceremony that I witnessed was conducted at a congress on forensic engineering. The thick proceedings volume from that meeting is full of papers on failures and their investigation, but the ring ceremony was curiously absent of references to those aspects of the profession. (In contrast, participants in the Canadian Iron Ring Ceremony are reminded of "the perversity of inanimate objects," which causes designs of all kinds to fail.)[30]

In its early years, one of the goals of the Order of the Engineer was "to have every engineering student become a wearer of the stainless steel ring upon graduation." Around 1980, Lloyd Chacey believed that 95 percent of the one million engineers in America would be affiliated with the Order of the Engineer in twenty-five years. In fact, it remains relatively uncommon to encounter an American engineer wearing the emblematic pinkie ring, although the number of engineers who are doing so is growing. In early 2010, there were in excess of 250 Links of the Order of the Engineer, and each year an estimated 10,000 engineers were reciting the "Obligation." Still, in the forty years following the initial Order of the Engineer ring ceremony, the cumulative number of American engineers who had recited the "Obligation" was probably no more than about 200,000. Over that same period, there were ten times that number of young college graduates entering the profession. In 2010 in Canada, where engineers number about one-tenth those in the United States, there were twenty-five Camps overseeing Iron Ring ceremonies, and approximately 350,000 Canadian engineers had recited the Kipling "Obligation" in English or in French in the eighty-five years since the inception of that tradition.[31]

But wherever they practice, the hundreds of thousands of engineers who wear the iron or steel ring on the little finger of their working hand do thereby constantly remind themselves and their colleagues of their dedication to their profession and of their obligation to society. Although most of these sons and daughters of Martha spend their lives working thoughtfully and carefully in the background on the furniture and machinery of civilization, they all wear their rings as constantly and proudly as Allan Bromley did his in Washington and as he later did in New Haven as dean of engineering at Yale.

Of course, not everyone is associated with a tradition like the Iron Ring. Those engineers who do not wear a ring are not necessarily any less mindful of their obligations to their profession or to society—or of the consequences of failure as they work on their designs. Ring or no, successful engineers know that it is always essential to concentrate on relevant failure possibilities if a design is not to fall victim to them. An engineer can imagine all sorts of wild and crazy modes of failure that might be irrelevant to a design's performance, but if a single critical one is overlooked and not designed against, it can be the one that causes a structure to collapse or brings a system to its knees. It is imperative that the realistic prospect of failure be kept in the forefront of every engineer's mind. Wearing an Iron Ring is one way of doing so.

Before, during, and after the Fall

The collapse of the Quebec cantilever span may be Canada's most famous bridge failure and the inspiration for its Iron Ring tradition, but the most infamous American bridge failure is undoubtedly that of the Tacoma Narrows. When completed in July 1940, the Tacoma Narrows Bridge had the third-longest suspended span in the world, surpassed only by the 3,500-foot main span of the structurally and architecturally revolutionary George Washington Bridge, completed in 1931, and the 4,200-foot central span of the iconic Golden Gate Bridge, which opened in 1937. Unlike those enduring structures, however, the Tacoma Narrows Bridge stood for only four months, its roadway torn apart in a 42-mile-per-hour wind on November 7, 1940. Because the suspended structure had revealed itself to be surprisingly flexible after its completion, its unexpected behavior was the subject of intense study and theorizing even before the collapse. When the nature of its motion changed dramatically from an up-and-down wave-like oscillation to a twisting movement on the fateful morning, a film crew rushed to the site to record it. The resulting footage of the dramatic last writhing moments and spectacular failure of the bridge's 2,800-foot-long center span became an immediate classic on the newsreel circuit, and it has since been shown to countless high school physics students as an oversimplified example of structural resonance—the natural phenomenon that causes wine glasses to break when a talented singer hits just the right high note.[1]

But film clips of the Tacoma Narrows failure are often shown with little context. The design of the bridge is seldom described in detail, nor are the features that made it vulnerable to a wind of not unexpected force. Without such context, the bridge appears to be an anomaly, and its failure a quirk of technology. In fact, the Tacoma Narrows Bridge was designed according to the state of the art of the 1930s. But other suspension bridges that were designed and built to the same standard in the second half of that decade, although also surprisingly flexible, did not collapse. Why the Tacoma Narrows did has been the subject of speculation, study, discussion, and debate among engineers and physicists for over seven decades, and there is little reason to believe that it will not continue to be. But, as with all structural failures, knowing the history of the bridge helps us to appreciate how and why it was designed the way it was, helps us to put its collapse in context, helps us to understand why explaining it is not a trivial matter, and helps us to comprehend why the consulting engineer most responsible for its design was found not to have done anything egregiously wrong.

Puget Sound is commonly associated with Washington State's largest city, Seattle, but in fact—via its passages, inlets, and bays—this large body of water reaches all the way down to Olympia, the state capital, more than sixty miles southwest of the better-known location. The city of Tacoma lies about midway between Seattle and Olympia, near a stretch of the sound known as the Narrows. Numerous toll ferries had long crisscrossed the sound and its arms, but a vehicular ferry at Tacoma was not established until 1929. The early-twentieth-century traveler who preferred to stay exclusively on land-based roads from Seattle to points west on the Olympic Peninsula had to drive the distance down to Olympia before heading back north. This was the situation when state planners began to talk seriously about a bridge at the Narrows. In 1929, the state legislature authorized a span at the location, literally the narrowest stretch of the sound. It would later be said that a Narrows bridge was also important for national defense, because it would provide a fixed crossing for reaching the navy yard at Bremerton, which is on the other side of the sound from Seattle and Tacoma. But "narrow" is a relative term, of course, for it is about a mile from the Tacoma side to the Gig Harbor side of the deep water with its strong tidal currents, conditions that make bridge building challenging and expensive. The crossing distance and considerable depth at the Narrows naturally required a major bridge to span it, and one that would leave enough horizontal and vertical clearance to allow ocean-going ships to pass safely under it.[2]

The state of bridge building at the time dictated that a bridge with a main span long enough to reach a good way across the Narrows had to be of the suspension or cantilever type. But after the collapse of the Quebec Bridge, the cantilever form had fallen out of favor for clear spans greater than about 1,200 feet, and so a suspension bridge was by default the only choice to be designed and built at Tacoma. It had been almost half a century since the Brooklyn Bridge, with a main suspended span of just under 1,600 feet, was completed in 1883. In the succeeding decades, longer-spanning suspension bridges had continued to be built, but each new record-setting span represented only an incremental increase, since the conservative-minded engineers of the time did not generally like to venture too far into the unknown. In 1929, the longest span in the world was that of the Ambassador Bridge between Detroit, Michigan, and Windsor, Ontario. Its suspended span measured 1,850 feet, representing only a 16 percent increase over the nearly fifty-year-old Brooklyn Bridge. But ambitious and confident engineers had two massive suspension bridges on the drawing board at the time. The bridge linking New York City and New Jersey across the Hudson River, later to be named the George Washington Bridge, would span 3,500 feet when it opened to traffic in 1931—a 90 percent increase, or nearly a doubling of the existing record—and the Golden Gate Bridge, linking San Francisco and California's Marin County when completed in 1937, would reach across 4,200 feet, a 20 percent increase over the George Washington. Given these daring precedents, albeit then still just plans on paper, it was not a stretch to expect a major suspension bridge with a span beyond any then in existence to be proposed to span the Narrows. Among the first designs offered was one by David Steinman, which called for a two-lane bridge deck atop a stiffening truss sixty feet wide and twenty-four feet deep. Had that bridge been built and properly maintained, it would likely still be standing today.[3]

Getting money to build Steinman's or any such bridge was no easy task. A Washington Toll Bridge Authority, modeled after successful California authorities that used toll revenue to repay construction indebtedness, was created in 1937. But in addition to the fiscal constraints of the Depression, potential sources of funding wondered whether the volume of traffic across a Tacoma Narrows bridge would generate sufficient toll revenue to repay a loan. The federal Public Works Administration (PWA) did approve a construction grant, but it was for less than half the projected cost of a bridge designed by Washington State engineers. And it came with the condition that the state highway department hire engineering consultants from the East, including the nearly legendary suspension-bridge designer Leon S. Moisseiff. Evidently, the consultants had convinced the PWA and the Reconstruction Finance Corporation, which was to provide a loan for the balance of the money needed to build the bridge, that it could be done for much less than the $11 million estimated by the state. The bridge would actually be built for $6.4 million.[4]

The preliminary design had been developed by Clark H. El-dridge, a 1918 graduate of Washington State College, who served as a bridge engineer with the Washington State Highway Department until 1939, when he was appointed bridge engineer for the toll bridge authority, where he also was in direct charge of design and construction for the Narrows bridge. The consultant Moisseiff, who had been involved with the design of the George Washington and Golden Gate bridges and who had taken part in a significant way in the design of virtually every major suspension bridge built in America since the Brooklyn Bridge, was critical of Eldridge's design, in part because it would have sloped down from the higher land on the Gig Harbor side to lower ground on the Tacoma side of the Narrows. Moisseiff, who was a strong proponent of aesthetic values in bridge design, convinced the toll bridge authority to adopt a level-roadway design, to locate the towers farther apart, and to make the profile of the bridge deck as slender as possible. The final design had a 2,800-foot main span, with a shallow deck stiffened not by the customary twenty-odd-foot-deep open steel trusswork but by eight-foot-deep solid steel plate girders. Since the Tacoma–Gig Harbor area was not then highly developed, in large part because there was no fixed crossing of the Narrows in the vicinity, the bridge was to carry only two lanes of traffic and have sidewalks less than five feet wide. The shallow, narrow road-way that stretched more than half a mile between towers thus gave the bridge an especially sleek and modern look.[5]

When engineers design structures beyond the limits of experience, they generally depart only a small amount at a time from what has proven to be successful. Until the George Washington Bridge was designed, a rule of thumb for bridge decks was to maintain a length-to-depth ratio—a measure of flexibility—of 150 or less. The design of the Tacoma Narrows Bridge resulted in a main-span length-to-depth ratio of 350, which was about the same as that of the George Washington Bridge. But that structure's eight

Washington State bridge engineer Clark H. Eldridge (1896–1990) and East
Coast consultant Leon S. Moisseiff (1872–1943) confer here on the roadway of
the Tacoma Narrows Bridge. Upon the insistent advice of Moisseiff, Eldridge's
initial design was modified to incorporate a more slender aesthetic, which
proved to be the underlying cause of the bridge's dramatic failure in November
1940. Had Eldridge's more traditional design been used, the bridge would most
likely still be standing.

traffic lanes and two wide sidewalks gave its deck a weight in excess of 31,000 pounds per foot of length, whereas the proposed Tacoma Narrows Bridge's very slender roadway resulted in an extremely light deck—less than 6,000 pounds per foot. Furthermore,
the literal narrowness of the deck gave the Tacoma Narrows Bridge
a length-to-width ratio of 72, significantly greater than the George
Washington's 33. Indeed, most existing suspension bridges had ra-
tios in the 30s, and of existing spans only the newly completed
Golden Gate Bridge had a length-to-width ratio as high as 45. The

ramifications of the extraordinary slenderness were not fully appreciated by the engineers most involved in the design of the Tacoma Narrows Bridge. Designers knew that the narrowness of the roadway would allow for considerable flexibility, and even before opening planners knew that the bridge deck could deflect about twenty feet horizontally in a wind of ninety miles per hour. But at the time that was the only kind of wind effect that was taken into account in the design of a suspension bridge.[6]

However, the extreme and unprecedented narrowness of the proposed bridge did not escape the notice of Theodore L. Condron, a septuagenarian consulting engineer who had been retained by the Reconstruction Finance Corporation to review the design and prepare a report declaring that a bridge constructed according to it would be a sound investment. Condron felt that the narrowness of the deck gave the bridge excessive flexibility, and he sought reassurances from engineers engaged in research on suspension bridges that the Tacoma Narrows design was a reasonable one. What he saw as a potentially fatal flaw in the design continued to bother Condron, but in the end he acceded to the authority, expertise, and clout of Leon Moisseiff, the chief consulting engineer (and effective designer) of the bridge that would be built across the Tacoma Narrows.[7]

Even before the bridge opened to traffic, it was clear that the slender deck was extremely flexible. Construction workers were said to chew on lemons to combat nausea while working on what would come to be nicknamed "Galloping Gertie." But engineers assured the public that the bridge's "bounce" was not dangerous. After all, several suspension bridges completed in 1939, the year before the Tacoma Narrows opened, also had a lively bounce to their roadway, but none of them appeared to be in danger of collapse. Nevertheless, once the Tacoma Narrows was officially opened to vehicles, cars driving over the bridge could see the traffic ahead of them rise and fall and thereby pass in and out of view as the roadway undulated up and down as if it were floating on waves on the sea. In an attempt to ameliorate the motion, the bridge was retrofitted with checking cables, but they did not fully eliminate the unanticipated behavior. Drivers and their passengers drove across the bridge just to experience the unusual phenomenon, and toll revenue surpassed all projections.[8]

The bridge behaved in this manner for about four months. During that time, the roadway had sometimes assumed the undulating form of a sine wave, but from side to side it had remained flat and not banked. On each of the bridge's sidewalks, the lampposts that rose and fell and tilted back and forth between the bridge towers had remained in the parallel vertical planes defined by the parabolic curves of the main suspension cables. This pattern continued until about ten o'clock on the structure's final morning, when as the vertical motion grew larger it suddenly took a novel twist. The large vertical motions of the bridge apparently had caused a cable band near mid-span to loosen and slip out of place, introducing an asymmetry into the forces restraining the deck. The bridge began to execute torsional oscillations, with the deck twisting about its centerline, as if sections of the roadway were aircraft alternately banking left and right. The lampposts now swayed in and out of the vertical plane of the suspender cables that connected the bridge deck to the main cables. The bridge was closed to traffic, and Professor Frederick Burt Farquharson, who had been watching the behavior of the real structure in conjunction with his experiments on a model of it at the University of Washington, rushed to the bridge to observe first-hand its new behavior. The camera crew set up on terra firma to capture the action on film.[9]

A lone automobile remained on the center span of the twisting bridge, abandoned by Leonard Coatsworth, a reporter who likely had been trying to drive across the moving roadway, perhaps to get a unique first-person story. In the film of the writhing bridge, the car is on the wrong side of the road, where it was thrown after its engine stalled. Even had it not stalled out, the heaving of the bridge deck would have made controlling the car a near-impossible task. After Coatsworth abandoned his vehicle, he became a sorry pedestrian who tried to scurry off the span but was repeatedly thrown about and had to resort to getting down on all fours and crawling along the rising and falling curb. He did crawl to safety, but with his hands and knees scraped and bloody.[10]

Professor Farquharson was captured on the same film, at times walking a bit like a drunken sailor along the relatively steady centerline, which remained nearly motionless even as the bridge deck alternately rose and dropped on either side or it. The steady centerline was what in physical mechanics is known as a nodal line. The engineering scientist Farquharson knew about such theoretical principles and applied them to aid his escape. He is believed to have risked his life trying unsuccessfully to rescue a small dog named Tubby that was trapped in the stalled car; the frightened cocker spaniel, who belonged to reporter Coatsworth's daughter, became the only life lost when the bridge eventually twisted apart and collapsed. The final minutes of the writhing of the Tacoma Narrows Bridge provided paradigmatic film footage of a real-life structural failure. Its scenes, as one reporter put it, "rank among the most dramatic and widely known images in science and engineering."[11]

The film of the bridge's collapse has received mixed reviews from engineers. Although it provides a rare glimpse of a full-scale structure exhibiting the extent of its flexibility, the film also provides incontrovertible evidence that engineers can make colossal mistakes—something of which few need to be reminded. Physicists, by contrast, seem to have had no reservations about replaying the film as if it were an endless loop, for it not only provides a visual real-world example of a dramatic mechanical phenomenon but also gives them an opportunity to assert that their theories can explain and predict behavior that at least some engineers clearly did not anticipate. It is perhaps no accident that the famous film of the infamous failure of the Tacoma Narrows Bridge was for a long time distributed not by an engineering organization but by the American Association of Physics Teachers. (Until the collapse of the World Trade Center twin towers in September 2001, which of course was videotaped from multiple angles by multiple witnesses and played over and over on television, the film of the collapse of the Tacoma Narrows Bridge stood alone as a reminder of how fragile even our largest structures can be.)[12]

On the evening of the failure, the consulting engineer and designer Moisseiff was in his office in New York City when he told an Associated Press reporter that he was "completely at a loss to explain the collapse." Moisseiff's junior partner, Frederick Lienhard, flew out to the West Coast that night to investigate the failure; Moisseiff planned to follow by train. The next day, he speculated that the accident was caused by "a peculiar wind condition." Ordinary conditions should not have brought down the structure, because, according to Moisseiff, it was "built according to regular rules." There was nothing unusual about the construction project, and no unusual materials were employed. However, he could not deny that even a light wind had been causing a "warping" of the roadway of the structure. That was why a model section of the bridge was being tested in a wind tunnel at the University of Washington.[13]

Charles E. Andrews, the construction engineer for the bridge, was not at a loss to explain what happened. In a radio interview on the day of the collapse he offered his opinion—which he stressed was only his opinion—that the collapse was attributable to the use of solid girders rather than trusses to stiffen the deck. According to Andrews, the girders "caused the bridge to flutter, more or less as a leaf does, in the wind. That set up a vibration that built up until the failure occurred." He observed that the Tacoma span was "the narrowest bridge in comparison to length, of any span, in the world." Andrews also compared the Tacoma Narrows to the Bronx-Whitestone Bridge, which had also been exhibiting undulations—though smaller ones—in the wind. However, he pointed out, that structure had a shorter span and was wider. In other words, its length-to-width ratio was smaller than that of the Tacoma Narrows. Andrews, who as construction engineer had lived with that structure, seemed to have understood the pertinent design features that doomed the failed bridge. Within two days of the collapse, the Associated Press reported that bridge engineer Eldridge "said that State highway engineers had protested against the design but that it was built as it was in the interests of economy." An editorial in the *New York Times* admitted that it was too early to state the cause of the accident, "though the construction engineers give it as their opinion that the introduction of stiffening plate or web girders in place of the older lattice or open girders was responsible."[14]

Some engineers are understandably reluctant to flaunt images of failure, but they do recognize that when a major suspension bridge so misbehaves that it collapses, the event stands as proof that the state of the art of long-span suspension bridge design left something to be desired. Such was obviously the case in November 1940, and so it may have been proper then to refer to the collapse as a "puzzle." An investigative committee appointed by the Federal Works Agency (FWA) was charged with finding a solution to that puzzle. The three-man committee comprised Othmar H. Ammann, the engineer responsible for the George Washington Bridge; Glenn B. Woodruff, engineer of design for the San Francisco–Oakland Bay Bridge; and Theodore von Kármán, the aeronautical engineer and director of the Guggenheim Aeronautical Laboratory at the California Institute of Technology. Early on, von Kármán attributed the cause of the Tacoma Narrows Bridge failure to something with which he was quite familiar—the shedding of vortices of air in a periodic manner, a phenomenon that created a wake known as a Kármán vortex street. In von Kármán's view, the wake reinforced structural oscillations that grew until the bridge deck could no longer hold together. The FWA report did not reflect von Kármán's convictions, however—probably because he was outvoted by bridge engineers—and declared that it was "very improbable that resonance with alternating vortices plays an important role in the oscillations of suspension bridges" and that the behavior of the Tacoma Narrows Bridge was probably due to "forced vibrations excited by random action of turbulent wind." Such uncertainties about the cause of the collapse can be found in the literature to this day. Most important, the committee found that aerodynamic forces, which were by then widely known in the growing aeronautical industry, were not taken into account in the design of the bridge deck, whose slender proportions made it behave more like a wing than a fuselage. Aerodynamic forces were not considered to be important by any engineers designing suspension bridges during the 1930s, and so the engineers of the Tacoma Narrows were absolved by the committee of any negligence or wrongdoing. But the failure certainly provided lessons for future bridge designers, not least of which was to be cognizant of technical developments in emerging fields of engineering and science.[15]

The fiftieth anniversary of the collapse of the Tacoma Narrows Bridge naturally occasioned recountings of the event. One article in the November 1990 issue of *Construction Today* declared that "in spite of official reports and 50 years of analysis, the violent failure . . . is still not entirely understood." Indeed, a great deal of research, writing, and debate had been focused on the cause of the Tacoma Narrows failure during the previous half-century, not only among engineers and engineering scientists but also among mathematicians and physicists who were not averse to applying themselves to the study of an artifactual phenomenon. Actually, although the bridge's failure may never be "entirely understood," for the actual structure no longer exists on which to test incontrovertibly any failure hypothesis, much progress toward an understanding has been made. And any lingering confusion may reflect a breakdown in communication among fields rather than a lack of research results.[16]

A cumulative index to publications of the American Society of Civil Engineers suggested that the society ignored the anniversary of the collapse. However, the December 1990 issue of *Civil Engineering*, the society's magazine, did carry a news item essentially announcing that a forthcoming paper in the *American Journal of Physics* would "spread the truth" about the cause of the infamous collapse. Since the paper was written by two engineers and promised to contrast an engineering explanation with those in physics textbooks, I looked forward to the journal issue, and I was not disappointed. Robert H. Scanlan, then a professor of civil engineering at the Johns Hopkins University, had studied the dynamics of structures over a long and distinguished career. As early as 1971 he took part in investigations of similarities between flutter in airfoils and in bridge decks—research that eventually led to a clarification of the nature of the Tacoma Narrows failure. The clarification managed, however, to escape widespread notice outside civil engineering circles until the later article written by Scanlan and a former student, K. Yusuf Billah, appeared in the physics journal. The inspiration to prepare a review article on the Tacoma Narrows Bridge collapse for a physics journal had come to Billah while he was browsing in a bookstore in the late 1980s. There he found himself examining three popular physics textbooks that "invoke inferences about the Tacoma Narrows episode that differ from present engineering understanding of the failure." A search of libraries and bookstores, he said, revealed the "ubiquitous presence of the Tacoma Narrows bridge failure in numerous other texts" (thirty are referenced in Billah and Scanlan's article). Almost all cited the bridge's behavior as an example of resonance.[17]

In their article, after quoting from several texts declaring resonance to have been the culprit, Billah and Scanlan admitted that the texts were qualitatively correct, but that they did not properly or quantitatively identify a source of periodic impulses—like the sound waves in a singer's glass-shattering voice—that could produce the resonance. It is generally implied that the wind itself is the source of the coincidence of frequencies, but no definite periodicity is typically associated with wind gusts or gales. Billah discovered several texts that actually resorted to von Kármán's explanation that shed vortices themselves created the periodic impulse that drove the bridge to destruction. To refute this theory, Billah and Scanlan calculated the frequency of the shed vortices in the 42-mile-per-hour wind that was blowing when the bridge collapsed. This frequency was about 1 cycle per second, which they noted is "wholly *out of synch*" with the 0.2-cycle-per-second torsional oscillations actually observed and measured by Farquharson.[18]

The engineers went on to reproduce a graph from a University of Washington Engineering Experiment Station Bulletin in which Farquharson reported on the aeroelastic behavior of a full-bridge model in a wind tunnel. The graph shows that, whereas the amplitude of vertical oscillation modes was self-limiting under increasing wind velocity, this was not the case for a torsional oscillation mode. According to the engineering-science model, as the Tacoma Narrows deck executed torsional oscillations, two kinds of vortices were shed. The first were those associated with the Kármán vortex street and having a frequency that did not coincide with the natural frequency of the bridge's motion. The second kind were complex vortices associated with the structural oscillation itself and having exactly its frequency—a type of vortex associated with the flutter wake that is created by a nonstream-lined body in an airstream. These motion-induced vortices, which predominate at large amplitudes of oscillation, apparently drove the bridge to destruction. Billah and Scanlan acknowledged a "'chicken-egg' dilemma: Did the vortices cause the motion or the motion cause the vortices?" They concluded that it was the latter. If there was resonance, it was complex and existed between the bridge's motion and the vortices produced by that motion itself.[19]

As Billah and Scanlan reported, the damping or shock-absorbing effects that would restrain oscillations in bridge decks change the direction of their effect at a particular wind speed, re-sulting in a phenomenon that allows the bridge to oscillate at in-creasingly large amplitudes, ultimately leading to structural col-lapse. Evidently the self-destructive torsional mode did not occur at Tacoma Narrows until a minor structural failure created asym-metrical conditions; once it was initiated it took only forty-five minutes to get completely out of control. Billah and Scanlan closed their paper on the Tacoma Narrows Bridge failure by observing that the "sensational photographs have made it an irresistible ped-agogical example." Furthermore, "because it lodges itself so easily in the memory, it is doubly important for educators to draw the correct lessons from this classic and sensational event." Indeed, a familiarity with case studies of failures is among the most effica-cious means of avoiding similar failures in the future. However, if the explanations accompanying the case studies are themselves flawed or misleading, they have the potential for doing more harm than good.[20]

Modern engineering rests heavily on mathematical and scien-tific foundations, and that is why the first two years of the engi-neering curriculum are dominated by mathematics and science courses. Eager and impatient engineering students often question the relevance of those courses to real engineering, and so the dis-cussion of real-world examples such as the oscillation and collapse of the Tacoma Narrows Bridge is especially fascinating to receptive and impressionable students. Teachers of engineering are repeatedly reminded of how difficult it is to break poor mathematics and science habits, especially those acquired in elementary courses that give preemptive explanations to dramatic engineering phenomena and failures. Yet in the Tacoma Narrows case study, mathematics and physics have been clearly behind the engineering science, for which they are properly prerequisite.

The juxtaposition of a simple, albeit retrospective, physical explanation and a complex engineering error of judgment and inadequacy of design has implications far beyond mere puzzle solving, for it contrasts the caricatures of the omniscient mathematician/scientist and the blundering engineer/designer. But there is no excuse for such oversimplification and stereotyping, whether explicit or implicit, in textbooks and classrooms. The collapse of the Tacoma Narrows Bridge will no doubt remain, as it should, an irresistible pedagogical example; it should not also remain a classic example of interdisciplinary hubris and conflict. To test any failure hypothesis fully and incontrovertibly, the bridge itself would have to be rebuilt, with all of its misaligned connections, loose joints, and other imperfections exactly as they were in the original. Since those realities of the actual construction can never be fully known, let alone faithfully replicated, any experiment to test a theory of the Tacoma Narrows failure can be open to criticism and challenge.

In the wake of the failure, engineering assessments demanded by the insurers of the Tacoma Narrows Bridge determined that no parts of its superstructure would be reusable in its replacement. Most of the center span was lying at the bottom of the Narrows—and remains there today, listed on the National Register of Historic Places and thus not to be disturbed even when the foundations for any new bridge are laid—and the side spans suffered such damage that they might as well have been underwater too. The towers, which had been designed to be pulled equally in each direction by the cables, were so asymmetrically pulled toward the nearby shore that they were permanently bent at their base and so, too, were not reusable. The demolition of the damaged structure was completed by mid-1943, with the salvaged steel contributing toward the war effort. Thus it is possible that molecules of the old bridge found their way into tanks and ships.[21]

The fact remained, however, that the first Tacoma Narrows Bridge did collapse in the wind, and no one wanted that to happen to any other bridge—especially not to a replacement bridge across the Narrows. The new bridge was to be built on the first one's foundations, the only part of the structure that was not irreparably damaged. This meant that the new bridge would have a span of the same length as the one that collapsed, and it was imperative that the new deck be able to resist not only the steady sideways push of the wind but also its stochastic bursts that might induce up-and-down or twisting motions. But at the time there was still no definitive theory of how the wind and structure interacted, so the design engineers had to rely on past positive experience with successful bridge decks and engineering lessons learned from failures. The design of a replacement Tacoma Narrows Bridge obviously had to confront the failure of the first. In particular, it had to ensure that the bridge deck would be stiff enough not to undulate or twist in the wind. This time there would be no joking about a Galloping Gertie, the replacement of which would be dubbed "Sturdy Gertie." The new bridge would have to demonstrate a contempt for the wind and a steadiness in it that would assure engineers and drivers alike that it could survive the storms expected to blow through the Narrows. Such assurances were achieved in large part by making the deck structure of the new bridge significantly wider and deeper than the original.[22]

The new bridge would be four rather than two lanes wide, reducing its length-to-width ratio by more than half, which placed it well below that of the Golden Gate Bridge and thus well within positive experience. Also, instead of stiffening the deck with plate girders, as had been done largely for aesthetic reasons on the first Tacoma Narrows Bridge, the replacement design resorted to the more conventional stiffening trusses. This not only gave the new bridge deck a smaller length-to-depth ratio, which brought it into conformity with existing bridges, but also opened up the structure to let wind pass through rather than buffet it and develop damaging vortices. The deep stiffening truss did not preserve the slender look of the original bridge, of course, but aesthetics was no longer an overarching design objective. What was obviously more important was that the bridge deck be stiff and steady in the wind—and look like it was.

Construction on the replacement bridge began in 1948 and was finished in 1950, making it the first major highway suspension bridge completed after the war. Other suspension bridges built in the 1950s included David Steinman's across the Mackinac Straits to connect the lower and upper peninsulas of Michigan. Completed in 1957 with a main span of 3,800 feet, it surpassed the George Washington Bridge as the country's second longest span and pushed the Tacoma Narrows down to fourth place in that category. But there is more than one way to state a bridge's size, and when measured from anchorage to anchorage the 8,614-foot-long Mackinac was the longest suspension bridge in the world—even though the Golden Gate Bridge had the longest main span. In 1964, the Verrazano-Narrows Bridge, connecting the New York City boroughs of Brooklyn and Staten Island over the narrowest part of New York Bay with a structure having a main span of 4,260 feet, pushed the Tacoma Narrows Bridge down to fifth place among all suspension bridges ever built in the United States.[23]

Early in 1951, the new Tacoma Narrows Bridge was tested by

several storms, in which wind speeds reached as high as seventy-five miles per hour. Since the first bridge had failed in a gale of only forty-two miles per hour, this baptism by wind provided proof of a sort that the new structure was indeed adequately designed. With a fixed crossing restored to the Narrows, and with the area sharing in the post-war growth that the entire nation was experiencing, the four lanes of the new bridge were well-used and appreciated by local drivers, commuters, vacationers, and sightseers alike. Indeed, with the development of the surrounding area that accompanies the completion of virtually any new bridge or road system, the fixed span and its approaches soon became busy with traffic. Toll revenue, which naturally increased with traffic volume, retired the loans needed to build the bridge, and tolls were removed from the Tacoma Narrows Bridge in 1965.[24]

"If you build it, they will come" may or may not be true of baseball fields, but it is almost always true of bridges. Although there have been a few "bridges to nowhere" that did not fulfill the dream, the rebuilt Tacoma Narrows Bridge was not one of them. Traffic volume continued to grow until the single crossing was no longer adequate to handle anticipated future growth. As early as the 1980s it was clear that an additional bridge would have to be constructed. When in the early 1990s state legislation encouraged the development of projects to improve Washington's transportation infrastructure, several proposals for a new bridge were brought forward, but authority to proceed was not forthcoming for another decade.[25]

Not surprisingly, given the growing popularity of the relatively new cable-stayed bridge type, such a structure was proposed to cross the Narrows. However, planners wisely realized that it would be in aesthetic conflict with the historic old suspension bridge, right next to which the new span was to be erected, and so they determined that the new bridge should also be a suspension bridge. In the United States, the most widely known side-by-side suspension bridges are perhaps the memorial bridges carrying traffic on Route I-295 across the Delaware River between New Jersey and Delaware and the Chesapeake Bay bridges located near Annapolis, Maryland. But the parallel Tacoma Narrows suspension bridge spans would be the largest pair in the world.[26]

Given the history of the original structure, designers needed to be sure that building a new bridge beside the existing Narrows bridge would not create some unanticipated wind-interaction effect that could threaten both structures. Carefully crafted twenty-four-foot-long aeroelastic models of the entire proposed and existing bridges were tested together in a large wind tunnel in Guelph, Ontario, where engineers had long tested models of skyscrapers designed to be built among a forest of existing tall buildings, where any change in the structural landscape can have adverse effects on the pattern of winds blowing through the canyons between them. No adverse wind effects on the side-by-side model bridges were observed. Still, because the Tacoma area is potentially subject to large earthquakes, designers also had to take seismic forces into account. In addition, since an earthquake might trigger a landslide along the bluffs beside the Narrows, the bridge's anchorages had to be designed to be immovable under such circumstances.[27]

The groundbreaking ceremony for another Tacoma Narrows Bridge was held in the fall of 2002. The complex project, which would involve the procurement of steel and the fabrication of deck segments in South Korea, demanded close coordination between design and construction if costs were to be kept under control. To accomplish this, the Washington State Department of Transportation, the owner of the bridge, issued a design/build contract to a joint venture of the San Francisco–based Bechtel Infrastructure Corporation and the Omaha-based Kiewit Pacific Company. Whereas the 1950 replacement Tacoma Narrows Bridge was built

at a cost of $14 million, the price tag on the new crossing was pro-
jected to be more than three-quarters of a billion dollars. The
design/build process would eventually bring the total project—
including work to upgrade the old bridge—in under budget at a
total cost of $735 million. Still, this represented more than a fifty-
fold increase in contemporary dollars over the first bridge. What a
difference a half-century makes![28]

But the passage of time, and the experience that comes with it,
does not necessarily mean that there are no new mistakes to be
made. Engineers, like everyone else, can always manage to find
new ways to miscalculate or overlook some detail. Murphy's Law
has never been repealed. In the case of the construction of the latest Tacoma Narrows Bridge, the prefabricated deck-and-truss segments from South Korea were piled up as deck cargo, as if they
were shipping containers, on ocean-going transport ships. When
one of the three top-heavy-looking ships was passing under the
old bridge to anchor just south of the new bridge under construction, one of the top-riding deck segments collided with the underside of the old bridge's stiffening truss. The damage was not nearly
as severe as the embarrassment. Evidently, someone had miscalcu-
lated the clearance under the old bridge at the time of passage. After recovering from the loss of face and making sure there was
enough clearance before any further maneuverings in the tidal
waters took place beneath the existing bridge, seamen moved the
new sections under it safely and construction workers hoisted
them into place on the new structure. Some were taken directly off
the ships from Korea, but most were first transferred to barges that
were floated into place beneath the awaiting cables. The order in
which the forty-foot-long segments were placed was crucial for
the proper loading of the suspension cables and towers. If, for example, too many deck sections were hung between the towers before counterbalancing ones were installed on the side spans, the

unbalanced force in the bridge's main cables might cause them to slip in the saddles over which they pass atop the towers or bend the towers permanently toward each other. Either consequence could alter the geometry and forces on the structure in ways that would invalidate all the careful calculations that had gone into its design.[29]

To the casual observer, the two side-by-side suspension bridges might appear to be identical twins. Indeed, their nearly equal-height towers are aligned beside each other and their truss-stiffened decks look very similar, but on closer inspection the towers and decks are significantly different structurally. The towers of the old bridge are made of steel, with three light horizontal double-X cross members above the roadway and a pair of large and heavy-looking single-X cross bracings stacked beneath. In contrast, the new bridge's towers are made of reinforced concrete, with only two horizontal braces above and one just below the roadway. Concrete was used for reasons of economy, but it was also used to avoid making the new bridge appear to be a carbon copy of the old. In an aesthetic nod to its neighbor, the new bridge's horizontal bracing has nonfunctional X-bracing cast into it.[30]

There are more structurally significant differences, however. Whereas the truss of the old bridge is mainly riveted together, that of the new is welded—not just beneath the main span but continuously throughout the entire bridge. And whereas the old bridge's deck truss is more than thirty feet deep, the new construction is about 20 percent shallower. Even so, the new bridge deck was designed to accommodate additional vehicle-traffic lanes or even a light-rail system beneath the roadway at some time in the future. (Suspension cables would have to be added to carry the additional load, but provisions for them have been made in the design of the towers.) The new deck is also wider, a reminder that traffic-lane width and shoulder standards have changed since 1950, something we are likely to be reminded of every time we drive over an older bridge.[31]

Like the 1950 structure, the new Tacoma Narrows Bridge has a total of four traffic lanes. When it opened in mid-2007, all eastbound traffic was routed across it, and all westbound across the old bridge. Making each bridge carry one-way traffic has obvious logistical and safety advantages. Tolls were reinstituted, but instead of being collected each way, thereby creating two bottlenecks, the relatively new (actually, now decades old) and enlightened method of doubling them and collecting them in only one direction is used. A round trip costs $3.00 cash or $1.75 by transponder, the fee structure obviously designed to encourage electronic payment and reduce toll-plaza overhead costs and traffic congestion. There are many more aspects to carrying vehicles efficiently across a waterway than just the steel-and-concrete bridges that are the most visible parts of our infrastructure. The steady pair of suspension bridges that now span the Tacoma Narrows attest to this, while at the same time they stand as a proud memorial to Washington State's resolve to recover from the engineering embarrassment of the fatally flawed first bridge, whose remains still rest deep beneath the water.[32]

The Tacoma Narrows is not the only bridge ever to have collapsed, of course, but it is unusual in that it did not involve any human fatalities. In this regard, the 1967 fall of Silver Bridge, which as we have seen claimed forty-six lives, is more typical. Most outright bridge failures do not follow months of obvious misbehavior. Rather, they follow times of much more subtle warning signs, which may or may not even be detected or detectable by the naked eye. As was the case with Silver Bridge, a hidden crack can grow ever so slowly—until it is of a critical length and suddenly gives way so fast that there is no time for those on its roadway to escape. While the cause of the Silver Bridge collapse was convincingly traced to a small imperfection in an eyebar that grew over the years into a crack, the cause of other catastrophic bridge failures can remain a mystery—or at least the subject of technical and legal debate—years after the fact. Failure is more often like a bad dream: It can progress silently behind our tightly closed eyelids, disturbing no one but ourselves, until suddenly it reaches an intolerable state, at which time something snaps and we are startled awake, screaming.

TEN

Legal Matters

Infamous bridge collapses are not the only devastating accidents to engender debate about their causes. In the immediate wake of virtually any catastrophic failure, especially one that results in loss of life, there is often a cacophony of speculation about exactly what happened and who was to blame. Unless there is a continuing action associated with the disaster, such as an ongoing oil leak in an environmentally sensitive location, the focus soon switches from the failure itself to its cause. Usually before too long, a favored theory emerges, which may or may not be correct but nonetheless becomes the focus of media reporting and public attention—until another favored theory displaces it. If a committee or commission is studying the failure, during the time it takes to collect background information, sift through evidence, and consider theories, the story of the failure tends to fade into the background and be eclipsed by more recent news, to emerge again mainly on anniversaries of the event. If the National Transportation Safety Board or some equivalent agency is involved, it may hold regular news conferences to provide updates on its progress, especially if some significant matter relating to design arises in the course of the investigation. The purpose of such public reporting is to disseminate even tentative lessons learned as soon as possible, so that structures or systems similar or even just analogous to the failed one may be looked into for the same design errors, lest they also fail. This kind of preemptive warning would be especially important, for example, if investigators discovered previously unidentified metal fatigue problems.

When a final report on a failure is eventually issued, it does not necessarily end the speculation, debate, or legal challenges. As we have seen, bridge failures that happened well over a century ago are still the subject of discussion as new sources of contemporary information and new tools of analysis combine to provide fresh insights into possible causes and scenarios. Because the original structure no longer stands, and because it cannot reasonably be expected to be reconstructed in its original form—not to mention the fact that it is highly unlikely that it could be loaded in exactly the same way as it was when the failure happened—any theory of how the failure occurred cannot be tested in any rigorous sense. That the speculation will continue should not, therefore, come as any surprise. This familiar way in which an investigation of a failure develops and evolves was played out in the wake of the 2007 collapse of a bridge in Minnesota.

Interstate Highway 35, which stretches from Laredo, Texas, to Duluth, Minnesota, is the main north-south artery running through the nation's heartland. Heading north, the road passes through San Antonio, Austin, and Waco, before dividing into east and west branches that pass, respectively, through Dallas and Fort Worth. After leaving that metropolitan area, the two routes rejoin to carry traffic on through Oklahoma City, Wichita, Kansas City, Des Moines, and then once again divide into an I-35E and I-35W pair that serves now the twin cities of St. Paul and Minneapolis, where the bridge collapse occurred.

Until a couple of years ago, I-35 evoked Texas for me. In the early 1970s, when we lived in Austin, it was the road my family and I took south to Laredo for a day's outing to the markets in Nuevo

Laredo, Mexico, or north to Dallas for a change. Known in the Austin area then as the Interregional Highway—or just the Inter-regional—I-35 was a wide manmade gash in the otherwise un-remarkable network of roads serving a slow and quiet town of modest proportions through which the lesser-known Colorado River flowed. Besides the river, the Balcones Escarpment—a visi-ble manifestation of the Balcones Fault—is the principal natural scar in the local topography and literally marks the transition be-tween the plains to the east and the Texas Hill Country to the west. Stretches of I-35 north of Austin more or less parallel the fault. When we drove home for holidays, we left I-35 in Oklahoma for interstates that ran northeast. When we visited Austin a few years ago, after a hiatus of three decades, we found traffic on I-35 through the greatly grown city to be directed to the upper level of a double-decked highway. Such an elevated roadway is essentially a series of bridge spans, of course, but we tend to forget that as we drive mile after mile over them. In all of my travels with my family on I-35 and other interstate highways, I seldom thought about the possi-bility of one of its bridge spans falling out from under us.

My primary point of reference for I-35 changed from Texas to Minnesota on August 1, 2007, when the bridge carrying I-35W over the Mississippi River at Minneapolis collapsed spontaneously dur-ing the evening rush hour. Thirteen people were killed and 145 more injured in the hundred or so vehicles that were on the bridge at the time of the structural failure that shocked the country and the world. Bridges carrying major highways are not supposed to fall apart so suddenly, though everyone knows that they occasion-ally do. It happened in 1967 to Silver Bridge, in 1983 to the I-95 bridge over the Mianus River in Connecticut, and in 1987 to the New York State Thruway (I-90) bridge over Schoharie Creek. When it happened in Minneapolis in 2007, drivers already on the bridge, not to mention their passengers, not only had no warning; they had no chance to do anything about the structure collapsing beneath them. It was remarkable that the death toll was not greater than it was. This can be attributed to how the bridge was constructed and how it fell. The triangular arrangements of steel members composing the trusswork designed to span the river and support the roadway from below effectively acted like crumple zones in an automobile, absorbing energy as they were being bent, twisted, crushed, and otherwise deformed as they fell and impacted the ground, the water, and the riverbed. This inadvertent design feature slowed the descent of the roadway to a survivable rate, transforming an otherwise hard, life-threatening landing into a relatively soft one for many of those in the vehicles traveling over the bridge at the critical moment, including a school bus full of youngsters, all of whom survived. But the design of the steel truss would also be the focus of forensic engineers and other accident investigators looking for an explanation of why the bridge failed after carrying traffic without incident for four decades.[1]

My wife and I were driving north on other interstate highways at the time of the Minneapolis accident. For most of our route from our home in Durham, North Carolina, to our summer retreat in Arrowsic, Maine, we follow I-95. Since we tend to talk rather than listen to the radio when we are driving long distances, I did not learn about the bridge collapse until the day after it had occurred. When we reached Maine, the news came in the form of telephone and e-mail messages from newspaper reporters and producers of radio and television shows requesting interviews. A call from the op-ed page of the *Los Angeles Times* prompted me to write that evening about bridge failures in general terms, including some historical examples, something I felt able to do even without knowing much about the Minneapolis bridge that was in the news.[2]

The media and the NTSB were not the only organizations interested in reporting on and investigating the Minneapolis accident. The state of Minnesota and its Department of Transportation certainly wanted to know what had happened. Minnesota and other states also wanted to know if any similar bridges might warrant immediate concern and inspection—and closing. The I-35W bridge was designed and constructed in the 1960s, and there were literally hundreds of other bridges of a similar design and age around the country. Was there something about the design itself that had caused the bridge to fail? Was there something about its being forty years old? Were there historical precedents?

The West Virginia Silver Bridge failure of 1967 had led to more stringent safety practices, including that of inspecting major highway bridges like the I-35W span in Minneapolis at least once every two years. The Minneapolis bridge, which had in fact been inspected annually during the seventeen years prior to its collapse, had thus passed scrutiny many times during its lifetime, and evidently no one had uncovered and reported a flaw or fault sufficiently concerning to warrant closing the structure to traffic. Only after the collapse revealed incontrovertibly that the span had had a problem did post-mortem inspections commence on the debris sitting in, above, and next to the Mississippi River. After a while, the wreckage would be moved to an NTSB-controlled site for further examination.[3]

But even before the wreckage was moved, steel bridges around the country were inspected more closely than they normally might have been, with some being closed as a precaution. And within a week of the accident, a consultant to the state of Minnesota, who had been poring over forty-year-old engineering documents, discovered a potential flaw in the bridge's design. He suspected that its gusset plates, the flat steel sheets of irregular polygonal shape to which beam- and column-like components were attached to form the joints of the truss, were overloaded. It also became clear early on that the overloading may have been caused by the weight of almost one hundred tons of gravel, along with heavy construction equipment staged in one of the closed lanes, awaiting use in a resurfacing operation on the bridge at the time of its collapse. This asymmetrical loading beyond what the bridge had been designed to support could have caused the underdesigned gusset plates to reach their limit of resistance. The NTSB would eventually come to a similar conclusion.[4]

In the meantime, without its major Mississippi River crossing, Minneapolis traffic was disrupted, to say the least. A replacement bridge was urgently needed, but bridges are not built overnight. Neither engineers nor citizens wished to see the same (or any) flawed design used for a new bridge, and so producing a new structure would require time spent in design as well as in construction itself. The most efficient approach would be to have a single firm or partnership design and build the bridge. With that in mind, the Minnesota Department of Transportation soon issued a request for bids, and within about ten weeks of the collapse a design-build contract was awarded to a joint venture consisting of Flatiron Constructors of Colorado, which describes itself as "one of the leading infrastructure contractors in North America," and Manson Construction of Florida, a leading marine building firm. The team promised to design and construct within less than fifteen months a replacement bridge that would last for a hundred years. But construction companies do not as a rule design bridges, and so Flatiron had taken on the Tallahassee, Florida–based firm of Figg Bridge Engineers as a partner to create the actual structural design.

The bidding process was expedited, but the contract that appeared to have been awarded on the basis of sketches and artist's renderings did not go unchallenged. Flatiron-Manson had not been the lowest bidder, and two competing bidders questioned the Department of Transportation's decision, which a judge would later describe as cloaked in secrecy. Eventually, however, the procurement process would be validated in the courts. In fact, the bids had been judged not solely on the basis of price but in a "best value selection process," in which factors relating to aesthetics and contractor reputation and performance record had been taken into account. Flatiron's design partner Figg was known for its striking, some would say artistic, designs. (The recently completed Penobscot Narrows Bridge and Observatory was one of the firm's designs.) And Flatiron, which had worked with Figg on other, similar bridge designs, had a well-established track record in building distinctive structures. Furthermore, Manson specialized in marine construction and so had the expertise to work with equipment floating on the waters of the Mississippi River. Thus there was a high probability that the team selected would deliver on its promises.[5]

The bridge that had collapsed was a steel deck-truss type that was described as "fracture critical," meaning that if one essential piece of steel in the assembly broke, the entire bridge would likely collapse. This kind of design would certainly not be chosen for the new bridge—nor would any kind of steel bridge. The Flatiron-Figg design was of the concrete box-girder type, which would be erected by the balanced-cantilever method, a construction process with which Flatiron had extensive experience. The hollow sections of the bridge would be cast on idle lanes of the closed interstate, trucked down to the riverbank, loaded on barges there, and then lifted by means of cranes to bridge level for placement. Once hoisted into place, each new section would be attached to its mate with intersectional epoxy adhesive and then pulled tight with interior steel cables that would tie all the sections together. Unlike the supporting cables of a suspension or cable-stayed bridge, the interior post-tensioning cables would not be visible in the finished structure, which would appear to have slender and gently arcing girders supported by graceful pillars sitting on the banks of the river. In fact, the replacement bridge would consist of twin parallel structures, one carrying northbound and the other southbound traffic. The main span of each was to be just over five hundred feet, well within the envelope of experience.[6]

Designing and building the new structure presented challenges and pressures beyond the technical. The Republican National Convention was scheduled to be held in Minneapolis in about a year, and so Minnesota's Republican governor wished to have a re-placement bridge as close as possible to completion by that time. This helped drive the accelerated design and construction sched-ule, which the Democratic mayor of Minneapolis and his political allies cautioned against. Furthermore, there was an ongoing de-bate in the twin cities area and throughout the state between those who supported more road building and those who favored the de-velopment of mass-transit projects, like light-rail lines. Indeed, there was pressure to incorporate a light-rail system into the new bridge. The compromise was to produce a bridge design that, al-though it would not carry rail lines initially, would not preclude the addition of a light-rail system at some future time.[7]

The basic design devised by Figg was presented to representa-tives of the community at a design charette that was held just six-teen days after the awarding of the contract. Laypeople were not expected to have input into the technical design of the structure, but they were called upon to participate in the selection of some of its aesthetic and cosmetic features and embellishments. The new bridge was presented as a "functional sculpture," incorporat-ing conceptually the theme of "arches—water—reflection." The arch theme tied the bridge into neighboring historic true-arch structures; water acknowledged that the bridge did cross a river; reflection referred not only to the fact that the image of the bridge would be inverted in the water but also that the structure should evoke the historical significance—that is, the failure—of the bridge's predecessor. Guided by the tripartite theme, the charette engaged the participants in helping to choose the shape of the bridge's piers, its color, and the nature of its lighting, railings, and abutments—none of which affected the underlying structural design. Not everyone liked the chosen design, which one citizen imagined was what people in the 1960s "envisioned the future would look like" and called the bridge "a throwback look to the future." The mayor of Minneapolis admitted that it was not a "groundbreaking design," and one woman declared that "they would've been better off doing a Calatrava bridge," referring to the Spanish-born engineer-architect whose striking signature designs had become much-coveted. But time was of the essence; traffic conditions were unbearable.[8]

Planners did allow time for some extra safety features to be added to the bridge, for it was being built in the wake of a colossal collapse. Additional engineering considerations for the new bridge involved the use of "smart bridge" technology. All bridges move: they vibrate under the wheels of passing cars and trucks; their roadways expand as the temperature rises and contract as it falls; they strain under a heavy accumulation of snow; they bend to the wind; and they shake when an earthquake occurs. The new bridge would be able to do all these things, of course, but it would differ from a conventional bridge in having had over three hundred sensors embedded in its concrete parts to record and transmit via fiber-optic cables data associated with expected motions and more. The data would be collected and analyzed at the nearby University of Minnesota to provide insight into the behavior of a real structure under real operating conditions and to serve related research programs. Some of the sensors would also provide early warnings should any element of the bridge behave unexpectedly, thereby providing a supplement to traditional inspection procedures and warnings of impending failure.[9]

In spite of the complications of politics, citizen involvement, winter construction, and the incorporation of smart technology, the new bridge was completed in slightly over eleven months.

Indeed, the new bridge was opened to traffic on September 18, 2008—just after the party convention but more than three months ahead of schedule. It is officially named the St. Anthony Falls Bridge, after the natural falls that are situated just upriver of the structure. A thirty-foot-tall abstract sculpture stands in the median at each end of the bridge to signal to drivers that they are entering onto a significant water crossing. These creative pylons consist of three closely spaced wavy columns that represent a "vertical interpretation of the universal symbol for water." They were cast with a special mix of concrete that reacts with airborne pollutants to keep them from staining the structure. Thus the pylons are expected to retain their gleaming white color without special maintenance.[10]

All the while the new bridge was being designed and constructed, forensic engineers were analyzing and deconstructing the nature and remains of the old. Early theories about the cause of the failure had included corrosion and metal fatigue, but the theory that the NTSB returned to repeatedly was the inadequate design of the gusset plates. They identified key plates that were half as thick as they should have been, but this alone might not have caused the bridge to fail, since it had stood for four decades with the same thin plates. Just because something is underdesigned does not mean that it cannot do its job; it all depends on how large a factor of safety was employed in the first place. If, for example, a structure is designed with a factor of safety of 3 throughout, but for whatever reason it was built with a critical component only half as strong as it should have been, then that particular part will still have a factor of safety of 1.5. In other words, it could still hold 50 percent more than its design load.

The NTSB evidently realized this, for the board's final report, which identified the underdesigned gusset plates as the most probable cause of the failure, also cited as a contributing factor the combined weight of the gravel and construction equipment that was parked on the bridge at the time of its collapse. This, in conjunction with weight that had already been incorporated into the bridge through earlier roadway upgrades, plus the rush-hour traffic at the time of the collapse, simply loaded the structure beyond what it could tolerate. The weak links were the gusset plates that were thinner and thereby weaker than they should have been.[11]

Such a scenario is not unprecedented in the history of bridges. Recall the case of the trussed-girder Dee Bridge that collapsed in 1847. The bridge had been carrying railroad trains safely, but there was concern that glowing cinders from a passing locomotive might fall on the wooden decking and set off a fire. To obviate that mode of failure, a heavy layer of ballast was added to the roadbed. The collapse occurred under the first train to enter onto the bridge after the ballast was laid down. The structure evidently could hold the additional ballast or the passing train, but not both simultaneously, and so it failed. Although there are obvious similarities with the Minneapolis bridge, it did not collapse under the first wave of traffic to use it after the heavy construction equipment and materials had been parked on it. There must have been other contributing factors, which is one reason the failure continues to be discussed and studied.

As persuasive to some as the gravel overload theory appeared to be in explaining the collapse of the Minneapolis bridge 160 years after that of the Dee, some engineers were not convinced. An investigation independent of the NTSB focused not on the gusset plates and traffic and construction loads alone but also on the corrosion factor. Since bridge spans do grow and shrink in length as they heat up and cool down, it is necessary to incorporate into their design some devices that are capable of allowing for such movement. These sometimes take the form of expansion joints, such as interleaved comb-like fittings, which carry traffic more or less smoothly over gaps that are wider in winter than they are in summer—and that are made audible by the impact and bounce of tires on them. Another change-accommodating device is the roller or rocker bearing, which allows one end of the bridge deck to move freely a small amount across the top of a pier, so that the structure neither pulls on the pier during periods of contraction nor pushes on it during periods of expansion. The bearing is also designed to prevent the connection to the pier from offering any resistance to the expansion and contraction of the bridge spans. If, however, the bearing becomes so corroded that it can no longer allow free movement, then the bridge might literally tear itself apart.[12]

August 1, 2007, was a very hot day in Minneapolis, and so thermal expansion should have caused the steel truss structure supporting the I-35W bridge deck to be close to if not at its longest length. But if its roller bearings were so corroded that they were frozen up, then the truss would not have been allowed to expand freely. This could have added additional compressive stresses to key members of the truss. Since the bridge also contained the thin gusset plates, some of which may already have been bent due to previous high-temperature loading conditions and so represented weak joints in the structure, they might have allowed lateral movement that could have gotten out of control. Thus, the combination of corroded bearings, bent gusset plates, high temperatures, and an overloaded roadway could have caused some truss members to fail by buckling and so have precipitated the collapse. This was among the arguments that lawyers representing victims' families were expected to make in court as they maintained that not only the designer but also the engineering company responsible for inspecting the bridge and the construction company responsible for parking its trucks and materials on the bridge were collectively responsible for the tragedy. Victims would eventually seek not only actual but also punitive damages from the companies they considered ultimately responsible for the collapse.[13]

Whatever the details of the failure, in 2008 Minnesota established a fund to compensate victims of the bridge collapse if they agreed not to sue the state. In an attempt to recover money paid out of the compensation fund, Minnesota subsequently initiated lawsuits of its own against the successor to the engineering firm that designed the bridge, claiming that the gusset plates were not sized properly, and against the contractor that placed the heavy materials and equipment on the bridge, claiming that the transportation department was not notified of the action. These were the same firms that were also being sued independently by survivors and families of the accident victims. As of the summer of 2009, however, the state had not initiated a suit against the firm that served as an engineering consultant for inspection of the bridge. This was confusing to observers, who noted that lawyers for the victims were accusing this firm of being negligent, in that over the course of three years before the collapse it had done "nothing to rectify the known problems with the bridge," thereby playing a "substantial" role in the failure.[14]

In its efforts to compensate victims and to expedite the construction of a replacement bridge, the state of Minnesota had paid out $37 million in the wake of the collapse. Subsequently, it also did sue the company that had been responsible for inspecting the structure. But the successor design firm appeared to have the deepest pockets. The original design of the Mississippi River crossing had been done by the firm of Sverdrup & Parcel Associates, whose successors in the meantime had been acquired in 1999 by the Jacobs Engineering Group. Lawyers for Jacobs first argued that their client was not liable because the design work was done forty years before the accident, and Minnesota law held that damage claims had to be filed within ten years of the damage and within two years after the discovery of it. The judge rejected this argument, however, stating that the suit against Jacobs was to recover money the state had paid the victims, which was an action within the two-year limit. Jacobs had in effect inherited the liability for alleged design errors made decades before it even acquired the original design firm. Two years after the failure, there were 121 pending lawsuits. Although cases were scheduled to be tried beginning in 2011, it was expected that there would be "a long legal aftermath" to the failure.[15]

As is often the case in the wake of a structural failure, theories explaining what happened to the Minneapolis bridge continued to multiply even as lawsuits were advancing through the justice system. One former construction company chief engineer had noticed something odd in a photograph that appeared on the Internet within a few hours of the collapse. According to the engineer, this photo of the collapsed bridge showed the concrete roadway all broken up and the steel trusswork upon which it once rested badly distorted. What caught his eye was an unpainted section of the top flange of a member of the upper part of the truss. He knew it had to be the top flange because his construction experience had taught him that this would be the part of the steel that would be embedded in the concrete of the roadway before the completed truss was painted in the field. This engineer also knew that the top flange of a steel beam should have had studs welded to it, so that when the concrete was poured it would have some means for gripping the steel and preventing the concrete and steel from moving in a shearing motion relative to each other. The materials would thus reinforce each other and make for a stiffer structure.[16]

In the photograph there were no shear studs visible on the unpainted flange, and there were no indications that any had been used. The engineer knew, again from his experience, that if shear studs were not welded directly to the flange, they would be welded through the metal forms used for the concrete. In either practice, the strength of the studs would have been tested by striking each of them with a hammer. A solidly anchored stud would either ring true or bend under the hammer blow. If there had been studs attached to the steel flange of the Minneapolis bridge, then they must have been sheared off when the bridge fell. However, the unpainted top flange in the photograph showed no sign of shear failures or of telltale welding-torch burn marks that should have been left by the process. In the absence of such marks, the engineer concluded that there never were any shear connectors on that section of the bridge, and that this omission may at least have contributed to the failure.[17]

The difficulty in reconciling competing theories of why a structure collapsed is largely due to the fact that strictly speaking they are just that—theories. Like scientific hypotheses, they should be tested to see which if any is valid. Unfortunately, when a structure collapses, it is obviously no longer available in its pre-collapsed form for rigorously testing any hypotheses. Rebuilding a collapsed structure to its pre-collapse condition to test it under failure-causing conditions is virtually impossible, since its parts are bent, broken, twisted, or lost. Using new materials to reconstruct the structure leaves any test of it open to the criticism that it is not the same structure as the failed one. Models (physical and computer) can be devised and built, but testing models is open to the criticism that they are just models, not the real thing.

When the NTSB issues a report, it typically identifies only a "most probable cause" of an accident, the wording being a recognition that a scientifically rigorous test of the hypothesis that condition A caused collapse B is unlikely to be achieved. Failure analyses of the wreckage of a structure can provide quite convincing and, in some cases, virtually definitive statements on the cause of a failure, but again, because they cannot be tested the statements remain generally unconfirmed hypotheses. An exception is the Kansas City Hyatt Regency walkways collapse of 1981, in which there was virtually incontrovertible visible evidence hanging from the roof that overloaded box beams were pulled over the washers and bolts of hanger rods. The structural detail was sufficiently simple and straightforward for investigators to draw essentially unassailable conclusions from its condition. Most structural collapses involve much more complex systems and interrelated factors, which by their nature allow for a larger number of possible failure theories. Legal arguments about causes of failures rely largely on expert testimony about hypothetical failure scenarios. The judge and jury in a case may be more persuaded by one hypothesis than by another, but a legal decision does not provide scientific proof that one hypothesis is more true than another. Legal arguments are not scientific or engineering proofs.

Even though the true cause of a particular structural failure may never be known with scientific or engineering certainty, the logic and reasoning embodied in a thoughtful failure analysis can hold invaluable lessons for improving the design of future structures. A particular hypothetical scenario may not have been the definitive cause of an actual failure, but it may potentially be the cause of a future one in a different structure. Incorporating that possible scenario into the design process is a way to obviate the feared failure. Failure analysis of structural collapses is as much about learning how to design better future structures as it is about understanding why something in the past has failed.

What was described as the "last major piece of litigation brought by victims" of the Minneapolis bridge collapse was settled three years after the accident. More than one hundred people had claimed that the URS Corporation, which Minnesota had retained as consultants to conduct inspections of the bridge structure, missed warning signs that the bridge was unsafe. In its defense, URS argued that it was not made aware of the structure's design vulnerabilities and so could not monitor them. The company, rather than engage in prolonged and expensive litigation, agreed to settle the case while admitting no fault. The settlement amounted to approximately $50 million, $1.5 million of which would go toward a memorial to those who had died in the collapse. In an attempt to recover some of their settlement costs, URS and the state of Minnesota continued to pursue claims against Jacobs Engineering.[18]

Regardless of what transpired inside and outside the courtroom, the failure of the I-35W bridge has become an inextricable part of the experience of crossing the Mississippi River at Minneapolis. The rusted steel truss that collapsed may have been replaced with a clean white sculpted concrete span, but few people crossing over the new bridge—whether everyday commuters or vacationers passing through on their way from Dallas to Duluth—will do so without being reminded, by the wavy pylons if necessary, of what occurred at that location on August 1, 2007. But those same drivers and passengers should also realize that what happened on a hot day one August is as rare as a March without snow in Minnesota. The fallen bridge had carried 140,000 vehicles a day—perhaps 200,000 people per day. In the forty years that the old bridge stood, well over a billion people may have crossed over it. Such large numbers do not diminish the tragedy that befell those who died or were injured in the accident, but it should reassure us that such incidents are indeed rare. And they can be expected to be even rarer in newer bridge designs, especially those created in the wake of failures.

Bridges, of course, are not the only things that fail under circumstances that are often euphemistically described as accidents. Airplanes, ships, and automobiles are also involved in accidents. Sometimes the failures are structural, such as when a wing or a tail falls off an airplane, leaving it without sufficient lift or means of control. Or when a ship spontaneously breaks in two, as happened to some Liberty ships during World War II. Or when an automobile has a blowout while traveling at high speed, causing the driver to lose control and the car to flip over. But these are failures of hardware, and we can eliminate or mitigate their risk by achieving a better understanding of the materials involved, by developing a more robust design, and by paying more attention to the maintenance and repair of the structure. Other failures are associated with software and with the way the components of a system interact with each other. These may or may not be described as accidents, depending upon one's point of view. The next chapter explores some of these softer aspects of failure.

ELEVEN

Back-Seat Designers

In the early days of digital computers, when by today's standards even large machines had little memory or storage space, many software designers made a decision that at the time appeared to be wise. Whenever a date was input, only the last two digits of the year were used. This not only saved time for data-entry clerks but also conserved space in a computer's limited storage system and simplified calculations involving things like the interest accrued in a bank account or the payment due on a credit card.

Abbreviating years was nothing new. Long before the advent of electronic computers, merchants and customers alike had become accustomed to the advantages of using two-digit years. Many a nineteenth-century printed business form already had its dateline partially completed: _____ ___, 18___. All the clerk had to do was fill in the month, the day, and the last two digits of the year. Ideally, forms were reordered as the new century neared, but if a large supply of the old ones remained at the fin de siècle, merchants could still use them by crossing out the "8" and writing in a three-digit year. Abbreviations or digits were commonly used to designate months, further saving time and graphite, even when there were no preprinted forms.

It has long been common to designate a date by means of digits and slashes, thus September 7, 1939, is rendered as 9/7/39. The order of the month and the day could vary, with Americans favoring month/day and Europeans day/month, but generally speaking everyone knew what continent they were on and so understood which convention was being used. With such familiar and efficient usage, what computer programmer or user could have been expected to doubt the wisdom of carrying the two-digit-year practice forward into the computer age? Those who did wonder what might happen at the beginning of a new century tended to dismiss the concern with the rationalization that by that time their programs would have long been superseded, or computer software and hardware would have evolved beyond the need to worry about such a trivial point.

Of course, as the year 2000 approached, it began to occur to just about everyone responsible for computerized data storage, accounting, and electronic transactions that a potential problem was looming. As predicted, computers and computer programming had evolved, but many old programs continued to be used beyond their implicit expiration date. These programs still calculated the elapsed time between two dates by simple subtraction, including the subtraction of one two-digit year from another. This had worked flawlessly throughout most of the last half of the twentieth century, but clearly there would be internal machine confusion at best when the digits 99 and 00, representing the years 1999 and 2000, were encountered. Subtracting the first from the second would result in a negative number of years, with which computers had not been programmed to deal.

Fixing what came to be known as the Year 2000 or Y2K problem would seem to be straightforward: programmers would simply go into the affected computer programs and change two-digit years to four-digit years in the algorithms. But many of the accounting programs were written in some of the oldest—and by then superseded—computer languages like the common business-oriented language, or COBOL. Because increasingly fewer younger-generation programmers were familiar with these languages, old hands who had retired were called back into service as consultants. Furthermore, there was some reason to fear that just changing two-digit years to four-digit years wherever they occurred would not take into account some subtle calculations within the computer code that clever programmers might have incorporated in the interests of further economy and speed. Finally, there was the very real concern that even if changes in computer programs were made correctly and completely, new errors would be introduced in the process. The only definitive test of the supposed fixes would come with the changing of the millennium, which most but far from all people believed would occur at the start of the year 2000. Nevertheless, mass-media anxiety grew as January 1, 2000, approached, and the world breathed a sigh of relief when it came and passed with no significant effect on the computerized business transactions upon which the global economy had become so dependent. Computer programs everywhere, it seemed, had been properly redesigned.[1]

The Y2K problem came to be viewed largely as a false alarm, and it quickly disappeared from public discussion. What many had feared was a design flaw with implications of earthshaking proportions proved to be much ado about nothing—at least from the perspective of the casual observer. In the end, the Y2K problem that caused so much anxiety at the close of the twentieth century proved to be, from the point of view of the vast majority of the public, a dud. And they soon forgot about it. Software designers should be less likely to forget the incident, for it had all the earmarks of a classic case study of what not to do in computer programming or in any other design situation—fail to foresee the future implications of a decision made today. Fortunately, the two-digit date glitch was identified in time to solve the problem. Unfortunately, not every potentially fatal design flaw is discovered before it does harm. In fact, before, during, and after the realization, popularization, and resolution of the Y2K problem, flawed designs of all kinds were discovered to be so only in retrospect—after an incontrovertible failure or near failure. One such example was the London Millennium Bridge, which shook so much when people first walked on it that it had to be closed after three days and redesigned.[2]

But what were the larger implications of the millennial computer problem for understanding failure and placing blame? After all, the use of two-digit years in computer programming might have led to a global financial meltdown, if the mass-media reporters and commentators were to be believed. What responsibility did computer programmers have to foresee the problem? Did they act unprofessionally or unethically in designing their programs with what proved to be a latent flaw that fortunately was caught and removed in time to avert danger? Were the programmers guilty of designing into their products what the automobile industry had so often been accused of doing—planned obsolescence?

When programmers were creating what proved to be limited-application if not flawed programs, they had to design within the constraints of very modest machine memory and high-priced computational time. The contraction of years to half their digits made perfect sense then. It not only saved bits and bytes of internal memory but also eliminated the unnecessary, time-consuming, and wasteful subtraction of the then-inconsequential digits "1" and "9" from each other every time a pair of years was involved. The pervasiveness of the Y2K problem among computer programs written in early programming languages indicates how designers tend to think alike when faced with a common challenge that has to be solved within common constraints. In other words, they tend to find (or find out about and adopt) similar resolutions because they are all immersed in what is referred to as the state of the art.

This groupthink is not conspiratorial; rather, it is a reflection of the fact that even the best designers recognize that there is no good reason for them to reject a clever design feature just because it is not of their own creation. By embracing the state-of-the-art solution, designers free up their minds to tackle other challenges, some of which they may meet in creative new ways that soon become embedded in an advanced state of the art.

With all the ifs, thens, ands, nots, nors, ors, and buts that early computer programmers had to juggle while seeking efficient and effective lines of code that worked individually and in concert with the whole, it is understandable that they welcomed the two-digit year as a quick and easy way to save space, time, and money. Of course, computer-programmers are not the only designers who face such challenges every time they work on a new design; designers of everything from airplanes to zeppelins have had to work within the state of the art of their specialization, and many if not most of them naturally embraced and used the consensus solutions to subproblems that they faced in common again and again. Each long-span cantilever or suspension bridge, for example, must be designed to support its own weight. If a rule of thumb or standard design procedure develops to ensure that the requirement is met, then designers tend to follow that procedure and get on with meeting challenges more unique to the latest structure on the drawing board. The Tacoma Narrows Bridge was designed in the Zeitgeist that emphasized slenderness over stiffness and that did not even imagine that wind could cause aerodynamic problems.

Choices are so much a part of design that some writers have gone so far as to define design *as* choice—or decision making. Indeed, the first sentence in the introduction of the first design textbook that I used in college read simply, "To design is to make decisions." Later, it added, "Decisions are always compromises." The straightforward and down-to-earth textbook evolved through many editions over the next half-century, and the eighth, published in 2006, told students that "mechanical design is a complex undertaking, requiring many skills." This may have been the publisher's way of justifying why "Shigley," as the book is known, had grown from 523 to 1,059 pages and also doubled in weight to five pounds in the interim. But no matter how complex and multiskilled design had purportedly become, it still remained fundamentally about making choices and decisions. That is why even though a suspension bridge is a suspension bridge is a suspension bridge, one designed by Othmar Ammann looks different from one by David Steinman. Both structures conform to the laws of physics and each has the unmistakable features that make it immediately recognizable as a suspension bridge, but the designer is in the details, which reflect choice. Ammann's bridge has his signature sleek, unencumbered high-arched open towers; Steinman's bridge shows off his taste for busier but far from baroque detailing.[3]

Any bridge form allows for plenty of unfettered self-expression on the part of its chief designer, but some aspects of a highway bridge over navigable water are dictated by what might be termed back-seat designers. The government charter to build the structure in the first place specifies the vertical and horizontal clearance that must be maintained for shipping lanes. Specifications regarding roadway lane width, guard rails, and lighting—both to illuminate the road from above and to provide warnings from the tower tops to low-flying aircraft and warnings from below to high-sailing watercraft—dictate certain aspects of design. Without these functional and safety regulations, there might be traffic chaos on the land, sea, and air.

Building codes constrain the design of habitable structures in a similar way. Smoke alarms and sprinkler systems; emergency lighting; means of egress in case of fire; structural resistance to earthquakes; window glass that can stand up to hurricane winds—

such safety features, where relevant, have become mandatory in new construction because of failures experienced in the old. In the 2001 attack on the Twin Towers of New York's World Trade Center, people were trapped on floors above the airplane impact sites when they could not get down past debris blocking the stairways. Lives were lost because alternate stairwell escape routes were located too close together in the core of the structure, and their proximity to each other meant that both suffered similar structural damage and consequent obstructions. In the wake of the tragedy, the city proposed new requirements for escape routes, making it mandatory that future designs incorporate more than two escape routes located at greater distances from each other, so that during a fire or other emergency at least one might remain passable, even if others were not. This rule change met with resistance, however, for it would require architects to design and owners to cede otherwise rentable prime office space over to non-revenue-producing stairwells.[4]

Few people would refer to the collapse of a skyscraper into which an airplane was deliberately flown as an accident. But normally we do use the word *accident* to describe the situation of an automobile crashing into a building, or into a lamppost, or into another automobile, or even into a pedestrian. Such accidents were feared from the earliest days of motorized vehicles. The development of a motor-car industry in Britain was inhibited by the requirement of horseless carriages that "a man carrying a red flag by day and a red lantern by night walk before any such contraption appearing on a British road." Early automobiles may have been feared, but they were little more than bare-bones machines in which to ride. They also provided a bone-shaking experience even though they could not go very fast, in part because of their limited power and in part because of the poor condition of roads designed—if they were in fact designed at all—for the more leisurely pace of horse-and-carriage transportation. Over time, as roads and roadsters improved, speed became an increasingly desirable and achievable pursuit—and a more dangerous one as well, especially in models whose steering mechanism retained the design features of that of a horse-drawn carriage. Of course, with higher speeds came higher-speed accidents as an inevitable by-product, and the number of highway deaths and injuries naturally began to rise.[5]

Early road accidents in the United Kingdom took a disproportionate toll on pedestrians and motorcycle riders, who together accounted for 85 percent of the fatalities in 1930. Road deaths of all kinds doubled over the period of the 1920s and 1930s, reaching almost 10,000 annually, then declined sharply during the war years, only to rise again in the postwar years, until around the mid-1960s, when safety features and practices brought a new steady decline. From a postwar high of about 8,000 annually in the mid-1960s, fatalities declined to under 4,000 in the mid-1990s.[6]

Of course, death on the highway is not a uniquely British—or American—phenomenon. Today, in emerging economies like China and India, where the number of roads and vehicles has increased rapidly, highway accidents are all too common. In India, deaths rose 40 percent over the five-year period ending in 2008, when there were almost 120,000 fatalities, with as many as 70 percent of them pedestrians. In addition to sharing the road with pedestrians, India's automobile drivers must negotiate around bicycles, scooters, and cows. As late as 2010, the only mandatory safety equipment for cars was seat belts. For additional safety features, like airbags or antilock brakes, buyers must pay extra. One observer described conditions on the road in India as chaotic at best: "On a recent day, a tractor hauling gravel was driving the wrong way, a milk truck stopped in the road so the driver could urinate and motorists swerved to avoid a bicycle cart full of wooden tables in the fast-traffic lane. Drivers chatted on mobile phones as they weaved across lanes and steered cars with stick shifts. Side mirrors were often turned in or nonexistent." The situation on roads in India has been attributed to a "total failure on the part of the government," which has been accused of inadequate planning and lax law enforcement, coupled with increasing numbers of vehicles, untrained operators, and drivers' licenses available for a bribe. In China the transportation system is more organized, with highway deaths having decreased for the decade ending in 2008, when there were 73,500 fatalities.[7]

China and India were recapitulating, in a way, developments that had occurred during the previous century in Britain and America. With the prosperity of the postwar period and the inception of the interstate highway system, increasing numbers of cars were being driven on America's roads. With more people in more cars clocking more miles at ever increasing speeds, accidents were bound to happen at increasing rates. The carnage on the highway began to reach epidemic proportions. Was there something about the way automobiles were designed that aggravated the situation? The consumer advocate Ralph Nader thought so, and he brought a heightened awareness of problems with automobiles to the attention of the public with his 1965 book *Unsafe at Any Speed: The Designed-In Dangers of the American Automobile,* whose preface began:

> For over half a century the automobile has brought death, injury, and the most inestimable sorrow and deprivation to millions of people. With Medea-like intensity, this mass trauma began rising sharply four years ago, reflecting new and unexpected ravages by the motor vehicle. A 1959 Department of Commerce report projected that 51,000 persons would be killed by automobiles in 1975. That figure will probably be reached in 1965, a decade ahead of schedule.

Nader was off in his prediction by only one year; the American highway death toll surpassed 51,000 in 1966.[8]

That same year, the cover of the paperback edition of *Unsafe at Any Speed* described the book as "the most explosive and influential best seller of the decade." It was indeed a scathing indictment of the automobile industry, which Nader convincingly characterized as concentrating on styling rather than on safety engineering. Summoning an overwhelming amount of evidence, he showed how Detroit automakers repeatedly neglected to incorporate safety and injury-mitigating equipment into their cars as standard features, even though the technology to do so was available. Selected safety features, such as seat belts and padded dashboards, were made available on new cars, but only as "safety options." The manufacturers argued that not enough car buyers selected these extras to warrant installing them on all vehicles. In other words, they were saying, there was no customer demand for safety, and car buyers did not want to pay for it. Yet, as Nader reported, the cost for producing a different automobile each model year, consisting overwhelmingly of cosmetic design changes like distinguishing chrome trim and tailfins, had been estimated at $700 per vehicle, which was passed on to the customer.[9]

Nader very convincingly condemned the industry for its failure to design car interiors to mitigate the effects of the so-called second collision, which occurs "when man meets car" immediately following the initial impact of one vehicle hitting another one or a stationary obstacle like a tree or a pole. In this second collision, the driver and passengers can be thrown against some sharp or unyielding part of the vehicle, with the result being injury or death. One recurring problem that was correctable by a design change was the steering column that could be forced back into the car's interior by a front-end primary impact, sometimes impaling the driver. Even in his sparsely illustrated book, Nader provided a page of conceptual drawings showing how steering columns could be redesigned to embody energy-absorbing and anti-penetration principles that would sacrifice the machinery rather than the human body.[10]

The automobile industry maintained that driver behavior and highway design—rather than vehicle design per se—were responsible for accidents and therefore their results. Nader clearly held the opposite view, and he made his case often and persuasively in his book. He also demonstrated with example after example how automobile manufacturers fended off federal legislation and regulation that would mandate some of the obviously beneficial design changes that would make cars safer for their occupants—and for pedestrians, whose injuries were often exacerbated by sharp and pointed hood ornaments and tailfins. Some state laws were successful in mitigating the effects of second collisions, requiring such things as safety glass, but it was very difficult to make federal headway—until the passage of a 1964 law that directed the General Services Administration (GSA), which was purchasing tens of thousands of vehicles per year for the federal fleet, to specify safety features that had to be on any car bought by the government. But the GSA safety requirements that survived political and industry lobbying pressures were weak, and it would take more than government purchases to move the auto industry to design for safety. Its record of not doing so continued to be unforgivable.[11]

Nader was not the only one who had become sensitized to the conditions on U.S. roads. He was able to make such a compelling case for improved design because he could draw upon statements like the one that opened a 1963 article in *American Engineer:* "It would be hard to imagine anything on such a large scale that seems quite as badly engineered as the American automobile. It is, in fact, probably a classic example of what engineering should *not* be." But it was the linkage between automobile accidents and public health issues that Nader returned to again and again. He described the work of one researcher as characterizing the goals of preventing disease and accidents as both being "fundamentally 'engineering' problems," and observing that "concentration on the hostile environment—the malaria swamp or the interior of a vehicle—is almost invariably more productive than trying to manipulate the behavior of people." And the problem was explicitly tied to failure: "From an engineering standpoint, when an accident injury occurs, it is a result of the failure of the technological components of the vehicle and the highway to adapt adequately to the driver's capacities and limitations. This failure is, above all, a challenge to professional engineering, which in its finest work has not hesitated to aim for total safety."[12]

Close on the heels of Nader's exposé, and after having conducted a three-year study of data on accidents of all kinds, a committee of the National Academy of Sciences reported on the toll of what it described as "the neglected disease of modern society":

> In 1965, 52 million accidental injuries killed 107,000, temporarily disabled over 10 million and permanently impaired 400,000 American citizens at a cost of approximately $18 billion. This neglected epidemic of modern society is the nation's most important environmental health problem. It is the leading cause of death in the first half of life's span.

Of the total deaths due to accidents of all kinds, almost half, or 49,000, were the result of motor-vehicle accidents. Nader had observed that whereas death due to "the old diseases," by which he meant such maladies as tuberculosis, pneumonia, and rheumatic fever, had diminished, death by automobile was on the rise. At the time of his writing, road accidents were the leading cause of death for the young and the leading cause overall. Of people hospitalized for injuries of all kinds, over one-third had been from automobile accidents. While the specific recommendations of the National

Academy report focused on medical response, the horrifyingly large numbers of accident victims, which were not unique to the report, understandably had already caught the attention of legislators. In particular, those concerned with motor-vehicle and highway safety had already been sensitized to the problem.[13] Whereas the death rate measured per vehicle miles driven was declining, Nader criticized the use of such statistics, noting that when compared with the total population, traffic fatalities had remained at about 25 per 100,000 since 1940. He stated that "what this means is that a motorist can expect to drive farther in any given year without being killed, but he is just as likely as in previous years to be killed within that year."[14]

The heightened awareness of the situation on the nation's roads led to passage of the 1966 National Traffic and Motor Vehicle Safety Act. This legislation provided for a "coordinated national safety program and establishment of safety standards for motor vehicles in interstate commerce." It led to the requirement that a number of accident-mitigating features be mandatory on automobiles, including seat belts for all passengers; steering columns that could absorb impact; padded dashboards; windshields made of safety glass; side-view mirrors; and doors that stayed latched in an accident. The legislation also led to the installation of guardrails, improved street lights, and improved median protection on highways. These measures had been a long time coming, but they alone would not solve all the nation's highway problems.[15]

In 1974, in the middle of an oil crisis, President Nixon signed legislation mandating a national maximum speed limit of 55 miles per hour. The stated intent of the law was to conserve fuel, which was not being used very efficiently at all at the higher speeds to which interstate-highway drivers had become accustomed. With the lower speed limit, annual highway deaths fell by almost 10,000 from an all-time high of about 55,000 in 1972. After the oil crisis passed, the speed limit was raised to 65, and highway deaths increased, reaching a relative high of just over 51,000 in 1980. But by 2009, the number of fatalities on U.S. highways had fallen to just under 34,000, or 1.16 deaths per one hundred million vehicle miles driven, the lowest rate since 1954 and down by 90 percent from the 1925 rate. The fatality rate dropped again in 2010, with the fewest people killed in traffic accidents in over sixty years. The reason for the steep decline in the most recent five-year period was "something of a mystery," but among the factors believed to be at play were that cars and trucks were increasingly being designed for driver and passenger survivability. Additional positive factors were public-service and law-enforcement campaigns for the increased use of seat belts, improved road design and engineering, and police crack-downs on drunk driving.[16]

The National Traffic and Motor Vehicle Safety Act of 1966 was landmark legislation that attempted to change the regulatory model in the United States from one of reaction to the highway accident toll to one of proaction. Prior to the legislation, the prevailing paradigm had been that drivers caused accidents, and so emphasis had been placed on changing driver behavior in order to avoid accidents in the first place. If evidence suggested that a particular vehicle design or design feature encouraged drivers to perform recklessly, that issue would be dealt with on a case-by-case basis. After the 1966 act, an epidemiological perspective prevailed, in which a vehicle was viewed as a risky environment in which driver and passengers were exposed to dangerous forces during a collision. Thus the objective became one of modifying those aspects of the vehicle that were the greatest potential cause of trauma.[17]

The 1966 act also had established the National Highway Traffic Safety Administration (NHTSA), whose responsibilities included investigating accidents and making safe-driving rules. Around 1970, when the agency's staff was approaching full strength, rule-making engineers outnumbered defects investigators by fifty-four to thirteen. A decade later there was parity, and afterward the investigators, who dictated recalls, dominated. Initially the rule-making engineers had developed and advocated for mandatory design safety features. However, the agency's focus changed following a 1974 amendment to the motor vehicle safety act requiring automakers "to remedy safety-related defects at no cost to consumers." With this provision, the rule-making focus of the agency changed to one of emphasizing vehicle recalls. Indeed, by the mid-1980s, the agency "supervised the recall of over half as many American motor vehicles as were sold new."[18]

Nevertheless, from its inception through the mid-1970s, the NHTSA did introduce safety regulations, some of which were vociferously opposed by vested interests. For example, the requirement that new automobiles sold after 1968 have factory-installed front-seat head restraints was challenged unsuccessfully in court by the Automotive Parts and Accessories Association, which feared that its members would lose "a lucrative market for head restraints as add-on equipment." The fact that headrests—as the restraint systems have come to be called, perhaps to deemphasize their origins and redirect the reason for their mandatory presence from accident safety to cruising comfort—had been available in the aftermarket emphasized that rule making regarding automobile safety features did not necessarily require a great deal of innovative mechanical design and development work by automobile manufacturers.[19]

Headrests were in fact developed to restrain the head from jerking backward in a rear-end collision. By limiting head motion, they reduced whiplash injuries of the neck. The efficacy of headrests having been established, and their presence required, automakers were forced to face a design problem whose solution can at best be considered a compromise choice made among competing and contradictory constraints. To meet its most basic function, an automobile headrest must be designed to restrain a driver's or passenger's head from being thrown too far back during a collision. Taken alone, this function can be met by devising a padded vertical extension of the seat back that is sufficiently strong to withstand a head's impact, while at the same time soft enough to cushion it. However, a variety of other functional and nonfunctional considerations complicate the challenge.

As we know from sitting in other than the front row in an auditorium or movie theater, different peoples' heads project different heights above the top of a seat back. To account for these differences in anatomy, fixed automobile headrests would have to extend from the top of the car seat and reach nearly to the roof. Some models of cars, vans, and trucks do have seat backs with integral headrests that do just this. Although this solution covers the range of passenger head levels, it does so at an aesthetic cost. Such vertically extended seat backs are generally out of proportion to the rest of a vehicle's interior—and clash with its predominant horizontal lines—and some interior designers must cringe at the monstrosities taking up space that otherwise had been so thoughtfully arranged. It is no wonder that the practical but objectionable solution of high seat backs has been generally rejected in favor of adjustable-height headrests.

Making a headrest adjustable in height means that the rest itself does not have to have a very large vertical dimension, as long as passengers remember to make the adjustment to the proper height. The size of the headrest is an important consideration when a more holistic or systemic view is taken of the various expectations of a vehicle's interior design. The headrest requirement stems from injury-preventing measures, but automakers have also confronted other safety issues. Among these is the desirability of providing the driver with maximum visibility in all directions. Overly large headrests can partially obscure the view out the right-side and rear windows of a car. Volvo ameliorated this problem somewhat by using headrests that are not solid upholstered masses but rather structures resembling a ladder that allow a Venetian-blind-like view of what is beyond them. Other automakers have opted for smallish headrests that presumably are adjusted to their proper height by whoever is sitting in the seat. But such low-profile rests all too often are left in their down position, thereby not serving their safety purpose, at least not for taller people.

Of course, vehicles are often occupied by more than the driver and front-seat passenger. Back-seat passengers can have a good deal of their view forward blocked by an overly wide or tall headrest. This can make riding in the back seat a claustrophobic experience, and the situation presents to the interior designer who acknowledges this problem the further challenge of finding a solution that is a compromise between the safety of the front-seat passenger and the quality of space of the rear-seat rider. Adjustable headrests with modest but safety-adequate dimensions do provide a reasonable balance between the competing objectives.

In one common situation front-passenger-seat headrests would seem out of place, however, and this is in taxicabs and limousines. Since the front passenger seat in such vehicles is typically vacant, the headrest atop it does nothing but obscure the view of the back-seat passenger, the customer. I had the enjoyment of once riding in a town car whose right-front headrest had been completely removed, and this opened up the view out the windshield in a most welcome and pleasurable way, even though it may have violated some regulatory dictate. Certainly it can be the case that the front passenger seat in a livery vehicle may be occupied, and in this situation the headrest should be put back into service. However, the general situation calls for the use of a headrest that can be folded down forward to a position below the top of the seat. Such a headrest can be at the ready to be flipped up when needed for a front-seat passenger but otherwise left folded down in consideration of the passenger riding in the back.

Seat belts and air bags are other mandated automobile safety features that present design problems whose solution can make the difference between success and failure, between survival and death in an accident. Early automobile seat belts were not much different from those found in the passenger seats of airplanes, but they evolved into more of a body harness, thereby preventing automobile occupants from having their upper body fold forward during a crash. Some of the more elaborate belt designs pulled away from the body when the car door was opened, and closed upon it when the door was closed. These kinds of devices presented some of their own design challenges, but generally speaking seat belts came to blend into the seats themselves or were concealed in interior finishes. The greatest challenge eventually was not so much designing the hardware but getting vehicle occupants to buckle up. Through the 1970s, seat belts were used by fewer than 15 percent of American drivers, but that number increased significantly as individual states passed laws requiring seat belt use, and in 2010 the compliance rate nationally reportedly stood at about 85 percent.[20]

Air bags were a more complicated design problem, because they needed to be concealed yet deployable at an appropriate speed and capable of inflating with an appropriate pressure to cushion a moving human body in a matter of milliseconds. However, the air-bag design appropriate for an average-size man not wearing a seat belt proved to be too much for children and even small adults, with some being killed by the very device that was supposed to save them. Deaths attributable to air bags fell but were not eliminated by advanced air-bag systems, which were capable of detecting the size of a seat's occupant and also taking into account whether a seat belt was being used. However, two out of three drivers killed in a crash in which the newer air bags—known as smart bags—were deployed were also wearing seat belts. Somehow there seemed to be a risk penalty being paid for clicking on the belt. The problem seemed to be that government standards continued to push design choices to favor the old paradigm, where the presumption was that seat belts were not being used.[21]

Another source of danger in automobiles is malfunctioning parts, whether mechanical or electrical. In the wake of a controversy involving sticking accelerator and brake pedals, in which Toyota eventually admitted to finding evidence that the pedals did indeed cause some fatal accidents, automakers and legislators reached a series of compromises regarding how the National Highway Traffic Safety Administration would establish and enforce standards relating to electronic components, something that had not been done before. Legislation is as much designed as hardware and software are, and the nineteenth-century philosopher Herbert Spencer is said to have observed that "British acts of Parliament all arose out of the failure of previously passed acts." The proposed Motor Safety Vehicle Act of 2010 was a model of compromise. Among the changes made in the bills before the House and Senate were to eliminate or extend deadlines, to give the secretary of transportation more discretion in rule making, and to allow for mitigation of the runaway acceleration problem rather than to absolutely prevent it. The deadline compromises were reached with the Alliance of Automobile Manufacturers, because the group claimed that further research had to be done. Among the organization's arguments for tempering the legislation with softer language was that "automakers cannot control all factors that could lead to unintended acceleration." They could not "prevent a shoe from going under a pedal" or "prevent people from putting in more than one floor mat." This is certainly true, of course, and it demonstrates the difficulty of designing a system that is not independent of its user. One aspect of the law that seemed to please both consumer advocates and auto manufacturers was the requirement that all new vehicles be equipped with devices—similar to the flight data recorders or "black boxes" used on airplanes—that would log data pertaining to the operation of the vehicle, thereby providing objective evidence of what might have caused an accident.[22]

In February 2011, the Department of Transportation released a report about Toyota acceleration problems that was based on tests and analysis done by engineers from the National Aeronautics and Space Administration, who know about computer-controlled propulsion systems. The report was described as offering "a measure of long-awaited vindication for the world's largest automaker." According to the secretary of transportation, commenting on the report, there was "no electronic-based cause for unintended high-speed acceleration in Toyotas. Period." But engineers tend to bristle at such absolute statements. Indeed, the NASA study said it was "not realistic" to attempt to prove that an electronic throttle control system could not under any circumstances produce unintended acceleration. Although NASA engineers had analyzed more than 280,000 lines of software code controlling the acceleration system, they could not conclude that it could not malfunction in the presence of, say, electromagnetic interference. A principal engineer was willing to go so far as to say that in the absence of finding malfunctioning electronics in the system, it was "unlikely" that it caused unintended accelerations, but he also allowed that "our detailed study can't say it's impossible." It is difficult to prove a negative under any circumstances, and this is why accident and failure investigations seldom lead to absolute certainty and finality about what may or may not have been a cause.[23]

Often in design a solution to a problem satisfies one constraint more completely than it does another, or pleases one party more than the other in an adversarial relationship. It behooves the designer to weigh competing objectives and come up with a solution that provides some happy medium that is, where necessary, also a fair compromise. Thus the bridge that must accommodate both vehicles and pedestrians can do so elegantly by elevating the walkway above the roadway, so that pedestrians can look off to either side without having their view obscured by traffic. Elevating the walkway also makes for a quieter and less hectic pedestrian experience. The Brooklyn Bridge is an excellent example of the kind of design solution that achieves this ideal while in no way impeding the flow of traffic. Indeed, this design feature provides the added bonus of a safer bridge, for vehicles seldom scale a roadside barrier ten or so feet high to injure pedestrians. The Calatrava-designed Alamillo Bridge in Seville, Spain, employs a similar but not so large vertical separation of pedestrian path and vehicle roadway. And the Gateshead Millennium Bridge, one of the most innovative and artful structures of its time, was designed with its pedestrian walkway separated from and on a higher level than its cycleway, thereby giving those who wish to take a leisurely stroll across the bridge the opportunity to do so in a safe and calming way.

Contrast these examples with a bridge like the Golden Gate, where the sidewalks are on a level with the roadway. Not only is there little spatial separation between strolling pedestrians and fast-moving vehicles (and their noise), but also it is virtually impossible for a pedestrian to gain an unobstructed view across the half-dozen lanes of busy traffic. This is not to say that the Golden Gate Bridge design totally fails the pedestrian, for it is possible for a person to walk on the bay side of the bridge and have the view and noise of its traffic blocked out of sight and mind by the spectacular view of downtown San Francisco and the widening bay in the distance. Elevating Golden Gate pedestrians above the traffic, at least without altering the basic architectural layout of the bridge, is a virtually impossible design (or redesign) task after the fact. Elevated outside walkways on the bridge would thicken its deck and thus alter its near-perfect proportions. Putting a walkway beneath the vehicular roadway would not eliminate the traffic noise and would put pedestrians inside the structure's stiffening truss, thereby introducing obstructions at regular intervals in both directions. The Benjamin Franklin Bridge across the Delaware River at Philadelphia does have its walkways elevated between the inside roadway and the outside train tracks, but the configuration gives the bridge deck a top-heavy look. (The Brooklyn Bridge does not suffer from this aesthetic flaw because its walkway is located in the center of the bridge and is bordered by lighter structural elements.)

Whether or where to locate a pedestrian walkway may seem like a small matter in the case of a big bridge, but it can be a decision fraught with aesthetic, economic, political, and safety issues. The final design of not a few bridges has been delayed or had its cost considerably increased because of the persistence of activist citizen groups in advocating for pedestrian and cycle access. In some cases, the final decision extended beyond the completion of the bridge, as was the case with Maine's Penobscot Narrows Bridge and Observatory. The new east-bay span of the San Francisco–Oakland Bay Bridge will have a fifteen-foot-wide pedestrian and cyclist path incorporated into it, and advocates were pushing for a similar amenity to the original west-bay span. The biggest obstacles to such improvements are the cost and how to fund it.[24]

Some high-level spans, like San Diego's Coronado Bridge, have been deliberately designed without pedestrian walkways in an attempt to discourage suicides. But this has not removed the risk entirely, for determined individuals have stopped their cars and rushed for the side barrier before anyone could react fast enough to prevent them from jumping over it. The Golden Gate Bridge has been among the most attractive to potential suicides, and great debates ensued when it was proposed to install a high fence atop its architecturally integral railing or a safety net extending out from beneath the deck. Proponents of such measures argue that saving lives is more important than preserving the aesthetic integrity of the iconic structure, which was naturally diametrically opposite the position of opponents to alterations. Designing a barrier—or anything—under such polarizing conditions is clearly a challenge that no matter how met is unlikely to make everyone happy. But meeting design challenges presented with competing objectives is more often than not the case in design of all kinds.

How an automobile designer treats a headrest or how a bridge designer handles the question of pedestrian access will always involve matters of choice influenced by a host of internal and external constraints, contradictions, and conflicts. If the choices made produce a part of the whole that appears and functions integral to it, the design will be heralded as a success and there will be little second guessing of what might have been. Consumers and users of designed things generally accept them as they are—as long as they are harmonious within themselves and with the relevant aesthetic. Designs that are not so readily accepted tend to be those that are clearly discordant within themselves, even though the source of the discord may not be easily articulated. In short, they are failures.

Failures can be obvious, as when a bridge collapses under rush-hour traffic. But not all failures reveal themselves so suddenly and incontrovertibly. In fact, failure is not necessarily something of which we can say with confidence that we will know it when we see it, even if it happens right before our eyes. Failures can be so subtle that they are hardly noticed, at least consciously, because they involve matters of judgment and taste, and so it can be debated if indeed they are even failures at all. This is the case with what some observers would deem aesthetic and functional failures, such as the arrangement of cables on a cable-stayed structure or the placement of a pedestrian walkway on a bridge. And the roots of many such failures can be traced back to the give-and-take of design by committee or even to the original conceptual design, which in some cases sets such a rigid agenda that ideas are locked in place. It need not be so.

The initial approach to solving many design challenges is the brainstorming session, in which a group of participants—engineers, inventors, designers, managers, entrepreneurs—is often implored not to fear failure. In an opening meeting to organize a team's approach to designing and building a robot that would compete in a national challenge, the high school instructor leading the team of seniors broke them into groups and encouraged them to brainstorm ideas, telling them, "No idea is a stupid idea. One idea that might never work could lead someone to think of another that will." The students were being admonished, as so many design teams have been, not to shoot down any idea, no matter how foolish it might at first seem to be; they were to let it live on the blackboard, the white board, the yellow pad, the foolscap, the Post-it note, or in the ether, for another day. The novice designers were taking their first step in the annual competition conceived by the inventor Dean Kamen, who came up with the idea for FIRST (For Inspiration and Recognition of Science and Technology), and their leader emphasized to the team that what they were embarking on was "engineering, not inventing." Lightbulb-inspired inventions would be welcome, of course, but what was really needed was for the team members to read books on robotics, look at websites relating to past FIRST competitions, and generally learn from the experience of others. According to the engineer-turned-teacher, "We need to keep looking at other ideas and find out what works best. Then make things better." In other words, they were to look for the shortcomings, not the obvious failures, of past efforts in order to make theirs more successful.[25]

Endless improvement is what engineering is all about. But the idea of improvement is intimately connected with the concept of failure, for if we do not recognize the shortcomings, the faults, the failings of something, we cannot even think of improving it. Improvement has its roots in failure. Virtually all inventions start out as improvements on something that already exists, and during their inventors' light-bulb moments they often appear to be the be all and end all. But when the glare of the lamp is shaded by the body of the inventor hunched over the workbench producing a prototype, or by that of the engineer at a workstation designing a production model and a factory in which to make it, or by the helmet of the welder installing the parts of a production line, or by the rose-colored glasses of the marketing manager calculating profit margins, or by the sunglasses of the salesman in the field representing the product, or by the visor of the accountant with a spreadsheet tallying profit and loss, the improvement begs for improvement of itself. As the FIRST instructor later quoted to his team, when they were buoyed by victories in a regional competition but facing stiff challenges in the next day's rounds, "Success is for the moment, and only that moment." It would not mean much if they failed the next day.[26]

Success is success, but that is all that it is. A successful design does not teach us anything beyond the fact that it works. We can ensure future successes by copying a successful design exactly, but that is often more easily said than done. A design is successful because no errors were made in its conception or planning or making or using: no corners were cut, no inadvertent substitutions were made in construction materials or method, no imperfect welds went undetected, no bolts were left untightened, no inspectors were deceitful or distracted, no maintenance was neglected. If any of these—and other—deviations from the original occurred in the design and fabrication and care of what might seem to be an exact copy of a successful design, then it is not that. And it may contain an unknown or unseen fatal flaw that the successful model did not—or the same fatal flaw that was in the successful artifact but had not yet been exposed by a failure. But even if we could do so, who wants just to make exact copies of successes? The world would be not only a boring place but also a limited place. If only successful bridges were copied exactly, then strictly speaking no longer ones could ever be designed and built. There would be no opportunity for poorer economies in need of greater spans to grow into richer ones. Great populations of people could be condemned to the condition into which they were born. Successful things serve best not as models but as motivators for improvement. Thomas Edison experienced the highs and lows of the quest for success in his search for a suitable material from which to make an effective lamp filament. He is famous for having tried thousands of possibilities, including, according to one account, "no fewer than 6,000 vegetable growths" from around the world, none of which he found to be quite right. When asked if he was discouraged by so many failures, he responded that they were not total failures, in that each experience was successful in teaching him that the thing tested did not work. It was as if Edison and his staff were following the sage advice that the nineteenth-century American educator Thomas H. Palmer gave to students in his rhyme,

> 'Tis a lesson you should heed,
> Try, try again.
> If at first your don't succeed,
> Try, try again.[27]

This sentiment has been echoed in many forms and contexts. Samuel Beckett, for example, in his parody *Worstward Ho* wrote: "Ever tried. Ever failed. No matter. Try again. Fail again. Fail better." By applying such dictums, however intended, Edison's laboratory did in time hit upon a successful filament material—carbonized cotton thread. Now, a light bulb going off in someone's head is the iconic representation for the inception of an idea—an "aha moment," an insight, a breakthrough, an invention, a design concept, an illumination, a putative success. But like the experimental lamps incorporating so many of the filament materials that Edison found unsuitable, even today's light bulbs can burn out immediately upon being switched on, as do most ideas. They go from the promise of success to the realization of failure in a flash of light. Still, the light bulb remains a universal symbol of the creative spark, and as short as the moment might be, it is something that is to be encouraged, cultivated, craved.[28]

A brainstorming session is not expected to produce finished plans or final products. The purpose of such an idea dump, or so we are told, is to get as many inventive thoughts shouted out and recorded in as short a time as possible, almost without thinking, as it were. The goal is to capture those light-bulb moments in the brief time between when they flare up and when they flicker out. At the end of a brainstorming session, when the conference room or project floor is knee-deep in light bulbs, the chore begins of cleaning up the mess—putting the good bulbs in one box and the failed in another, with the former to be saved and the latter to be disposed of. But how are the good distinguished from the bad at such an early stage in a project? They are not yet screwed into a network of power and purpose that can test them. Even if they were, how long would those that burned bright continue to do so? Typically frosted, the bulbs are not transparent and their filaments are not manifest. Their insides and thus their promise are hidden from view. Some designers might believe that the good can be distinguished from the bad by shaking the bulbs, listening for a failed filament. But some bulbs are three-way, with more than one filament. Even though one may be broken, rendering the three-way function inoperable, that does not mean that the bulb may not still be usable. In which box does it go? And what of the blown bulbs, do they not also have something to teach us? Knowing why they blew out can be more instructive than knowing that others did not.

As useful as failure may be as a teaching tool or an inspirational device, it is obviously undesirable when lives are at stake. No designer or engineer wants his or her machine or structure to fail catastrophically when it is being used. That is why engineers especially think about failure when designing their gadgets and sys-

268   TO FORGIVE DESIGN

tems. If engineers do not properly anticipate how failure might occur in what they are designing, then they may not think to incorporate into the design a defense against its happening. But no matter how thoroughly a design might be vetted for possible ways in which it can fail, there can be no assurance that some theretofore unimagined interaction among parts under some unforeseen circumstances might produce some new and undesirable effect. There can be no assurance that the scale or ambition of the new device, structure, or system has not taken it into an unknown realm, where physical phenomena previously of no great significance begin to dominate performance and thereby precipitate failure.

Bridges are paradigmatic structures whose design and construction serve as models of those creative and practical processes that have enabled civilization to flourish amid a constant interplay between failure and success. That this interplay is as old as civilization and its sine qua non, engineering, is embodied in stories of the kind told by Herodotus and Caesar—stories of successes built on the lessons from failures, by the way—in their accounts of campaigns into enemy lands over deliberately designed bridges and by the obviously engineered artifacts, like the Pyramids and Stonehenge, that speak of design even when there is no written record of the accomplishment. The art of bridge building continues, of course, through the present day, and plans for future bridges are on the minds of engineers who dream of spanning the Bering Strait—thereby connecting North America and Asia—and the Strait of Gibraltar—to connect Europe and Africa. Throughout history, bridge building has been one of the great human endeavors, and it promises to continue to be so indefinitely, in both fact and metaphor. After all, what were missions to the Moon and what will be manned missions to the outer planets other than bridge-building projects of great accomplishment and ambition?

Other than those failures intended by design—such as the

blown fuse that prevents an electrical circuit from becoming over-loaded—failures might be said to be accidents. But are accidents of design caused by the designers themselves or by the drivers of the vehicles by which the design ideas are transported to proto-types and further down the manufacturing highway to produc-tion? Or is there something inherently wrong with the design pro-cess itself? Do failures, like highway crashes, continue to happen because of the risk-taking and inattentiveness of the person at the wheel, or do they continue to happen because the design, manu-facturing, and operating environment is inherently unsafe at any speed?

TWELVE

Houston, *You* Have a Problem

All things, and especially systems in which people interact with things, fail because they are the products of human endeavor, which means that they are naturally, necessarily, and sometimes notoriously flawed. Some technical flaws may be as innocuous as personality flaws are in the individuals who introduce them, but just as we come to ignore the annoying habits, tics, and traits of people with whom we become familiar and comfortable, so do we tend to ignore the idiosyncrasies of technological parts and systems that we work within and rely upon. Knowing that nothing is perfect, we do not expect perfection from our technology, and so we grant a pass to the recalcitrant valve or the stubborn switch, giving it a second chance to perform its function. When it does, we tend next time it sticks or wiggles to give it a third and fourth chance. Eventually expecting it not to work on the first, second, third, or even fourth time we try it, we try it even more times. We take such behavior as part of the thing's mechanical or electrical personality, accepting it as good-naturedly as we do the fits and foibles of someone who has become a dear friend, or at least a familiar face. This is the way it was with imperfect O-rings and insulating foam on space shuttles. Their imperfections became just part of their personalities.

Few catastrophic failures have been as public in their occurrence and as lingering in their impression as the explosion of the space shuttle *Challenger* only seventy-three seconds into its January 1986 flight. The launch had been much anticipated, in large part because among the shuttle crew was the first teacher to go into space, and from orbit Christa McAuliffe was scheduled to broadcast lessons to schoolchildren in classrooms across America. In the wake of the accident, videotape of the failed mission was televised over and over, with increasing attention paid to what at first seemed like a small and inconsequential leak emanating from one of the booster-rocket joints. As we eventually learned, the jet of hot gas weakened a nearby bracket securing the external fuel tank, which in turn broke away from the shuttle proper and wreaked the havoc that ended in the explosion. The gas leak was eventually attributed to a faulty O-ring seal, a design detail that engineers had previously identified as a weak link in the system after noting erosion of the rubber-like rings following some earlier flights. Indeed, because cold weather was believed to aggravate the leaking condition and because it was expected to be uncommonly cold on the morning of the scheduled launch of *Challenger,* engineers had recommended against proceeding. All this is now well known, as is the fact that managers responsible for giving the go-ahead to launch did just that. According to Richard Feynman, writing in an appendix to the report of the presidential commission charged with investigating the accident, "The O-rings of the Solid Rocket Boosters were not designed to erode. Erosion was a clue that something was wrong." According to Feynman, failure preceded catastrophe.[1]

The presidential commission had conducted its investigation over the course of three months, and testimony taken during that time filled 15,000 pages. Hearings were conducted in both closed and open sessions, the latter of which were televised. Among the highlights of those meetings was the tabletop experiment conducted by Feynman, who had immersed a small O-ring in a cup of ice water on the table before him. When he took the chilled object out of the glass and undid the clamp that had held it in the form of a dog bone, it did not regain its original circular shape immediately, showing that it had lost some of its resilience. Similarly, a cold O-ring improperly seated in the booster-rocket joint might not have sealed it properly. Feynman's persuasive demonstration and his reasoning regarding the wisdom of the engineers' warning and the recklessness of the managers' ignoring it to give the go-ahead for the launch have been an enduring image and lasting moment of the failure investigation. His testimony was effectively embodied in the commission's conclusion that "the decision to launch the *Challenger* was flawed."[2]

The design of the O-ring seal itself was flawed, of course, but respecting its limitations the engineers felt comfortable relying upon it at higher temperatures. Had their warnings about launching in excessively cold weather been heeded, the accident probably would not have happened. The management structure and organizational culture both at NASA and at Morton Thiokol, the manufacturer of the booster rockets, were ultimately blamed for the failure. A congressional committee that held its own hearings on the accident also found that the technical problem with the O-rings had been identified in time to halt the launch, but that "meeting flight schedules and cutting cost were given a higher priority than flight safety."[3]

The legacy of the *Challenger* explosion should have been a more cooperative and respectful partnership between engineers and managers, not only at NASA and its contractors but also in technological organizations generally. Unfortunately, what may be inherent weaknesses in human nature seem ultimately to have prevailed. After a thirty-two-month hiatus, shuttle missions resumed, ostensibly with improved design features and management practices. Eighty-seven successful missions followed, but upon the launch of *Columbia* in January 2003 a piece of insulating foam broke away from the external fuel tank and struck the leading edge of the shuttle's left wing, impacting an area designed to shield the structure against the extreme heat of reentry. Engineers expressed concern about the damage that might have been done, and while the mission was proceeding carried out some impromptu experiments simulating the foam hitting the wing. In spite of engineers' calls for detailed remote photography or shuttle crew inspection of the damage, and in the face of lingering concerns and uncertainty about the wisdom of allowing the shuttle to reenter the atmosphere in a damaged state, managers downplayed the effects of the foam strike and cleared *Columbia* for reentry. It soon became evident that the craft's thermal-shield integrity had indeed been compromised, with hot gases finding their way into the wing's structural interior. Images of the ship's disintegration over Texas provided sad data for another investigation and yet another opportunity for gathering lessons learned, which proved to be strikingly similar to those presumably learned after the *Challenger* disaster. According to the report of the Columbia Accident Investigation Board, "the NASA organizational culture had as much to do with this accident as the foam." Two years after that report was issued, seven members of the group of twenty-five that had been monitoring NASA's efforts toward improving shuttle safety declared that managers and officials of the agency had to break their "cycle of smugness substituting for knowledge."[4]

Space shuttle flights resumed again two and a half years after the *Columbia* disaster. The program may have continued safely for the rest of its scheduled missions, but it was not only NASA and allied organizations that should have learned much longer lasting lessons from the failures of *Challenger* and *Columbia*. Wherever managers—especially those who do not have a technical background—interact with engineers, there is the potential for a dysfunctional relationship. When something goes wrong in such an environment, the first instinct is often to blame the engineers and their designs, but on further probing, the root cause of a failure is often something much more subtle and complex than just a sticky mechanical component, a shorted electrical switch, or a flawed design detail. It almost invariably involves conflicts between risk-averse engineers and risk-tolerant managers. The 2010 oil spill in the Gulf of Mexico—really a succession of tragic events and failures, a tragedy of errors—is a more recent but already classic case study that was widely reported on during the seemingly endless period over which it evolved.

In many if not most of its issues during the late spring and early summer of 2010, the *New York Times* carried related stories filling two, three, and four pages, each headed "Spill in the Gulf." On a typical day, at least one of the stories was continued from the paper's front page. Maps showing areas of oil concentration and projected movement of the oil slick began to appear within weeks of the accident. After about a month, maps showing the results of the latest aerial survey and the daily forecast for where the oil would reach next were augmented by a box headed "The Latest on the Oil Spill," summarizing the status of the situation and highlighting developments in the response. On June 2, six weeks after the explosion that announced the failure, the status box was headed "Day 42: The Latest on the Oil Spill," the newspaper evidently conceding that this was something that might go on for some time, and that the *Times* was going to count the days and continue to provide summaries of important developments in attempts to cap the gushing well, efforts to contain the spill, progress on the cleanup, and the environmental and political fallout.[5]

Television networks, too, closely followed developments in the Gulf. On CNN, Anderson Cooper hosted a seemingly nightly show on the spill, often broadcasting from the Gulf Coast, as did many other news outlets. The technological failure that eventually entangled managers, bureaucrats, politicians, and pundits in its web was clearly news for both wide and narrow audiences alike. The trade publication *Engineering News-Record (ENR)*, the weekly magazine of the construction industry, carried stories emphasizing the technical, industrial, and regulatory aspects of the spill. *ENR* has a long and rich tradition reaching back into the nineteenth century of reporting on failures, especially of large structures and systems. In early June 2010, when the oil leak occurring 5,000 feet below the surface of the Gulf had been ongoing for about six weeks—and when patience and tempers were wearing thin, especially along the Gulf Coast and in the nation's capital—an *ENR* editorial entitled "The Gulf Oil-Spill Disaster Is Engineering's Shame" concluded with the threatening observation that "the Gulf of Mexico always will be remembered as the place where engineering prestige dipped to a new low in an age known for disasters as much as for progress."[6]

Readers have always been quick to respond to stories in *ENR*, often offering early explanations of failures that the previous issue of the magazine had just reported upon. In the case of the editorial on the oil spill, there were letters to the editor galore. One termed it "an insult and assault on the engineering profession" and called for a retraction. Another reader accused the magazine of holding the position that "EVERYTHING can be blamed on engineers," giving hyperbolic examples of users holding them responsible for failing to design skillets in which egg yolks never break and roads in which potholes never develop. Except when employed as unrealistic examples, such perfect products are not expected to come out of the design, manufacturing, and construction process. In fact, the design of a skillet, road, or offshore oil rig is complicated not only by the laws of nature and the limitations of materials and processes but also by the interaction of engineers with others in an office, firm, corporation, or government agency. Another letter to the editor put it thus: "What is seldom appreciated is that once management, technologically illiterate, intervenes and changes the design, it owns the consequences." The reader saw "no difference in the scenario played out in the Gulf and a host of other significant engineering disasters, such as the Challenger and Columbia fiascoes."[7]

Obviously space shuttles and oil rigs are vastly different in hardware and function, and no one would expect a space shuttle to resemble an oil-drilling rig or expect a rig to fly. However, each of these large pieces of advanced machinery was but the most visible and tangible manifestation of an enormously complex technological system. The shuttle did not conceive of itself: that required the vision of NASA and supportive politicians; it did not fund itself: that needed congressional appropriations; it did not design itself: that required a team of engineers; it did not plan its own missions: that was usually overseen by scientists and managers; it did not ready itself for launch: that was done by technicians; it did not launch itself: that needed booster rockets and an external fuel tank, designed by people; it did not operate itself: that was done by computers programmed by software engineers; it did not land itself: that was accomplished by astronauts gliding it home. The shuttle program did not organize and run itself: that took managers and administrators. So neither does an oil rig conceive of, invest in, design, or operate itself. That is the work of corporations, engineers, and technicians. Nor does an oil rig choose the time, place, or method by which it will drill as deep as tens of thousands of feet beneath the water's surface. That decision falls to scientists, engineers, and managers—with at least the tacit approval of company officers and boards of directors. All complex technology is inseparably coupled to an equally complex team of people and systems of people who should interact with one another as smoothly and with as clear a purpose as a set of well-meshed gears. What occurred in the Gulf of Mexico in the spring and summer of 2010 was at least as much a failure of the human machinery as of the mechanical devices involved.[8]

The *Deepwater Horizon* was a half-billion-dollar drilling rig built in South Korea by Hyundai Heavy Industries. It was completed in 2001, and at the time of the accident was owned and operated by Transocean Limited, a Swiss company that described itself as the world's largest offshore oil and gas drilling contractor. Reported to have been "one of the most sophisticated drilling rigs on the planet," *Deepwater Horizon* was of the semisubmersible type, which means that it had a variable draft, riding almost forty feet higher in the water when being moved from site to site than when partially submerged for increased stability while engaged in drilling operations. The rig was immense in size, at about 400 feet long, 250 feet wide, and 135 feet tall. It was dynamically positioned, so that by means of thrusters it could stay in place over a drill hole even in waves almost thirty feet high and winds of sixty knots. *Deepwater Horizon* was designed to drill in water as deep as 8,000 feet and to drill down 30,000 feet—over five miles—to reach oil deposits. In addition to carrying the machinery and equipment necessary to accomplish this task, it had 130 berths to accommodate the needed crew. At the time of the accident, the rig was licensed to the global corporation named BP—formerly known as British Petroleum—and was operating over the oil company's *Macondo* oil well, about forty-two miles southeast of Louisiana's Mississippi Delta. On April 20, 2010, an explosion occurred and the ensuing fire burned for two days, until the rig sank in the mile of water. Eleven workers were killed, and seventeen were injured.[9]

Rescue and recovery efforts naturally dominated the early response and reporting, but it was soon clear that the disaster was not over when the fire was out. A five-mile-long oil slick developed, indicating that there was a leak somewhere in the just-completed well system. Like all wells drilled at such water depths, where the hydrostatic pressure is about 150 times what it is at the surface and the well pressure is naturally even greater, that beneath the *Deepwater Horizon* had been fitted with a remote-controlled

In April 2010, an explosion on the oil-drilling rig *Deepwater Horizon* set off a fire that led to the vessel's sinking in about 5,000 feet of water in the Gulf of Mexico. The accident caused the rupture of the pipe string rising from a complex mechanical and hydraulic device known as a blowout preventer sitting on the bottom of the Gulf. The failure of the preventer to close off the well under emergency conditions allowed oil to leak virtually uncontrollably for months.

blowout preventer, in this case a fifty-four-foot-tall, mechanically complex set of pipes, hoses, flanges, valves, pistons, and rams designed to cut off flow from a well that was running out of control. On the night of the explosion, the preventer did not function as designed, and as oil subsequently flowed uncontrollably out of the wellhead and its broken pipes, plans were devised to try to fix the leak. But because of the depth involved, the work had to be done remotely. Within days, the Coast Guard approved the operation of remote-controlled underwater robots to prod and repair the recalcitrant preventer, but they were not able to do so. Early estimates of the rate of oil leaking uncontrollably from the well were about 1,000 barrels (that is, 42,000 gallons) per day.[10]

In addition to the submarine robots, cameras were operating near the wellhead, and they showed the area littered with severed pipes from which oil and natural gas were escaping in three places. Estimates of the oil leak were upped to 5,000 barrels per day. With the robots unable to stop the leaks, BP initiated the first of what would prove to be a confusing number of exotic attempts to capture at least some of the oil. Several of these efforts might even have been described as comic if the situation were not so tragic. Seventeen days after the explosion, a forty-foot-tall, ninety-eight-ton welded steel "containment dome" was lowered over one part of the broken pipe. The rectangular inverted-box-like structure and attached mile-long string of pipe were designed to funnel the oil up to surface ships that could capture it and thereby keep it from polluting the water and coastal regions. Those working to solve the problem anticipated that ice might form in the riser pipe, and so warm surface water was circulated in an outer pipe. Unfortunately, the design did not take into account the fact that the cold water temperatures surrounding the source of the leak would also create what was variously described as a "slush of frozen hydrocarbons" and "an ice-like slush of gas and water," which did clog the assembly and render it ineffective. Any design is only as good as its designers' imagination, and in this case that was not good enough. This embarrassment was soon followed by a number of others, and the oil-industry jargon by which they would be characterized made them the object of some ridicule.[11]

Two days after the containment-dome failure, BP announced that it might attempt to stop the flow with a "junk shot," in which a variety of objects like parts of shredded tires, golf balls, plastic cubes, and knotted ropes would be pumped into the blowout preventer, with engineers hoping that the debris would clog it up the way children's toys can a toilet. The technique, which was more technically described as injecting a "bridging agent" into the piping system, had in the past stopped gushing wells around the world. One experienced well tamer, who had employed junk shots successfully in the Kuwait oil fields following the Persian Gulf War of 1991, described the technique as blending scientific principles and historical experience with what had worked before. However, a junk shot had not been tried in such deep undersea conditions as existed in the Gulf, and the great pressure with which the oil and gas were escaping kept any effective amount of the junk from getting lodged in the nooks and crannies of the piping system.[12]

A battle of the pressures ensued, whereby engineers attempted to use downward pressure to overcome the 13,000 pounds per square inch of upward pressure pushing the oil and gas out from the well. One such effort involved a so-called top kill, in which heavy drilling mud—a slurry consisting of clay, water, and barium sulfate, which is literally heavy—was pumped into the leaking well in the hope that the accumulated weight of mud filling the pipe would be sufficient to overcome the well pressure. If that were to work, the well would then be sealed permanently with cement. Simultaneously, a smaller containment dome, resembling an inverted funnel, was being prepared to cover the blowout preventer and direct the oil leaking from it into a riser that would take it to a surface vessel for capture or treatment. One great disadvantage of the approach was that it would make it difficult if not impossible for the undersea robots to access parts of the preventer and try again to get a critical part of it unstuck. Nevertheless, the so-called top hat would eventually be placed over the nonfunctioning blowout preventer, but only after its damaged riser was cut off with some difficulty. In the meantime, a four-inch-diameter pipe was successfully inserted into the broken twenty-one-inch well pipe, catching some of the crude oil before it could enter the water. This "first success" worked to some degree, diverting the captured oil up to a ship that could process it. However, a larger pipe could not easily be employed because the pressure from the well might eject it, resulting in no oil at all being captured. After almost four weeks of embarrassments, BP was relieved that something had worked as planned and described the string of failures as having been "all part of reinventing technology." It was a process of "learning, re-configuring, doing it again." BP was wringing success out of failure. However, unlike designs that are tested in the friendly environment of a company laboratory where failures were private lessons, these designs made under organizational pressure were being rushed into field testing under the glare of media scrutiny. It was not a comfortable setting for engineers.[13]

Almost three months into the spill, BP tried a new tactic. The loose-fitting containment cap that each day had been funneling as many as 25,000 barrels of oil to a ship on the surface, where it was burned off, was removed. As the company announced, this allowed virtually the full discharge from the well to escape into the water, but the short-term damage caused by this measure was expected to be justified by the complete well-capping that was expected to follow. After the containment cap had been removed, robots undid each of the six fifty-two-pound bolts holding a large fitting onto the flange atop the blowout preventer, thus providing a clean connection point. To the exposed flange surface robots attached a new containment system that when fully closed would, BP hoped, capture completely the flow of oil. The fitting was successfully attached, but before all of its valves could be closed a battery of pressure tests had to be performed to establish the condition of the well. If there was structural damage to it, closing it off fully might create such high internal pressures that the oil it contained might burst out somewhere other than at the wellhead. Unfortunately, the testing procedure had to be postponed so that additional analysis recommended by the secretary of energy and others could be done first.[14]

Over the next week or so, as valves were methodically closed on the ninety-ton capping system, pressure began to rise, indicating that the well casing was not obviously leaking. When some seeping oil and possible methane gas were found near the well, the government and BP disagreed on what course to follow, with the government wanting the oil company to be at the ready to open valves in order to release pressure on the well. BP, which had finally capped the well, wished to leave it that way until it could develop further options. If oil did again gush from the capped well, the plan was to put in place a looser-fitting cap that had been held in reserve should the tight-fitting one not have worked. This would funnel much of the oil up to surface ships, where it could be burned off or otherwise processed. In fact, the battle of the pressures had evolved into a battle of public relations and politics. The president of a public-relations firm, commenting on the ongoing debate, was reported to have said of the government and oil company players that "they want to project being on the same team, but they have different end results that benefit each." Those calling the shots in Washington wanted to lower the risk of worsening the situation, while BP did not want to retreat from its long-sought achievement of stemming the leak, since the amount of oil released would determine the fine levied on the company. A proper balance of risks and benefits is always especially difficult to achieve when different parties push and pull from each end of a loose alliance. As long as there were "undetermined anomalies" present on the seafloor, the government only extended its permission day-by-day for BP to keep the cap tightly on and required the company to do careful monitoring of the well pressure. But finally, after nearly three months of leaking oil, the announcement came that the flow from the well was stopped.[15]

With that enormous milestone accomplished, BP could begin the process of sealing the well. This involved pumping drilling mud into the well from above, followed by cement, an operation known as a "static kill," with the process sometimes referred to as "bullheading." After much media anticipation of this procedure, it finally began about three weeks after the capping took place, and then over the course of about eight hours 2,300 barrels of heavy drilling mud were pumped through the blowout preventer and into the well casing. When the procedure was halted to verify through pressure readings that no new leaks had developed, a "static position" appeared to have been achieved. At this point, Thad Allen, the retired Coast Guard admiral in charge of the government-controlled response to the oil spill, expressed optimism that the end was near. Within days, cement was poured on top of the mud, but because there remained uncertainty as to whether the cement would fill the annulus between the inner and outer well pipes, two relief wells continued to be drilled. When one of them intersected the main well, the final cementing of the notorious well from the bottom up would be able to proceed. Once that "bottom kill" procedure was completed, the well could "never be un-killed." What by then was an anticlimactic milestone was preceded by a shift of media coverage of the oil spill "from engineering to the environment and the economy." But even as the shift was taking place, CNN's Anderson Cooper wondered if the spill had been "overblown by politicians and scientists and media," including himself.[16]

Now that the well was capped and officials had tentatively established how much oil had been spilled, the administration in Washington announced that only about 25 percent of the leaked oil remained to be dealt with. Almost 75 percent had been contained, collected, cleaned, skimmed, evaporated, burned off, dissolved, dispersed, or had otherwise somehow disappeared. Whatever oil remained in the sea was expected to break down relatively quickly. Tropical storms that had interrupted operations were credited with having helped disperse some of the unwanted oil into droplets, which oil-eating bacteria already had or would quickly cause to biodegrade. Oil that had reached Gulf Coast beaches, often in the form of tar balls, had been hand-collected by work crews and disposed of like garbage. Other oil was buried in sand or sediments and therefore out of sight. But some skeptical environmental scientists disagreed with the government's estimate of how much oil remained to be dealt with and of its potential impact. Scientists claimed that independent research found that as much as 80 percent of the oil from the well was "still lurking below the surface," some having settled to the seafloor and rendering it toxic to fish-spawning areas. In other words, the true environmental cost of the spill might continue to accumulate well into the future.[17]

Still, even as the media covered every angle of the story, crews tested various schemes to stop the gushing and collect the spreading oil, and officials debated strategies—and almost since the leak itself was known to exist—the two relief wells were being drilled as a sort of effort of last resort. The idea of a relief well was to intersect the main well at about 18,000 feet, deep enough beneath the seabed so that drilling mud could be injected upward into it. In this "bottom kill" procedure, the rising column of heavy mud would eventually weigh so much that it would exert more pressure down on the oil than it would up on the mud. However, the process of drilling a relief well was always known to be a slow one, and in the *Deepwater Horizon* case such a well had not been expected to be completed until mid-August, four months after the explosion. One of the reasons that relief wells traditionally took so long to complete was that as the relief-well shaft drew closer to the original well, the drilling had to be stopped at regular intervals so that the drill bit could be hauled up and a magnetic ranging device sent down to get precise measurements on the location of the target. Since the well pipe was made of steel—a magnetic material—it influenced the electromagnetic field of the Earth, and the extent of the effect provided information on the pipe's location. Had the old technique been used in the Gulf catastrophe, as the relief well reached a depth of almost 17,000 feet, just 1,000 feet short of the bottom of the target well, it would have taken about seventeen hours to complete the cycle of retracting the drill bit, pushing the sensing device to the bottom of the relief well, taking measurements, retrieving the device, and reinserting the bit to resume drilling. However, the technology had advanced considerably by 2010, and detecting devices shared space in the relief well with the drill bit, so that the long trip up to the surface and back was no longer necessary. Still, according to Admiral Allen, progress on the wells had to be "very, very slow because they have to be very exact." In fact, even with such deliberateness, the relief well could miss its target. Indeed, in working to plug a leak off the Australian coast the previous year, crews made five attempts before successfully intersecting the leaking well with the relief well. For a while, there had been some optimism that the relief well in the Gulf of Mexico might even be completed before August, but the original projection of mid-August proved to be more realistic. Anyway, with the top kill appearing to have been successful, the pressure of speed was somewhat off the relief wells. Indeed, weeks after the successful static kill, "government scientists were still studying the test results to determine the precise procedures to be followed in completing the relief well."[18]

Throughout the entire ordeal, there had been no shortage of ideas from inventors from outside the oil industry and government laboratories about how to stop the leak. However, many ordinary citizens who felt that they had a workable scheme were frustrated that they could not get through to the White House or the Department of Energy, from which some significant direction was believed to be coming, or to BP. The company did have a place on its website for offering suggestions for stopping the leak, but it provided limited space in which to describe an idea. Eight weeks after the explosion, BP had received in excess of 80,000 suggestions for dealing with the spill. The rate at which ideas were coming in increased as the oil washed up on the shore. At least one idea, from the engineer and inventor Willard Wattenburg, made it to the desk of Secretary of Energy Steven Chu, no doubt because the proposer had been responsible for capping over five hundred oil wells that the retreating Iraqis had left burning in Kuwait in 1991. At the time, scientists estimated that it would take five years to put those fires out and cap the wells, but Wattenburg accomplished the task in seven months. His idea for stopping the Gulf of Mexico leak was to drop hundreds of tons of steel balls into the well. If the balls were of sufficient size, he reasoned, their weight would cause them to fall through the rushing oil and eventually clog up the pipe enough to finish the job with drilling mud. The secretary wrote back to Wattenburg that the Department of Energy had considered such a scheme but that it had complications, which remained unnamed. Other recommendations, such as using explosives—even a nuclear device—to blow up the well and fuse it closed also had complications, which were more obvious.[19]

All the while, ideas for capping the leaking well had continued to be coordinated from a command center in Houston, where video monitors showed conditions undersea. The space commandeered was that normally used to coordinate operations during hurricanes, and it was staffed with employees not only from BP but also from Transocean and Halliburton, who were working on the rig when the explosion occurred, and even from competitors like Exxon Mobil. One BP vice president explained how the five hundred or so engineers, technicians, and support staff were dealing with the unprecedented situation: "We are designing every option to be successful, and we are planning for it failing." This explains why so many disparate schemes were being developed at the same time. About three weeks into the crisis, Secretary Chu, who was almost invariably described by President Obama and the press as a Nobel-prize-winning physicist, had been variously reported as being "dispatched" and "lent" by the White House "to help come up with solutions" to the problem of the oil leak. If conditions leading up to the accident itself were complicated by the engineer/manager dichotomy, then the leak-stopping efforts were complicated further by the involvement of scientists and politicians, and the cleanup complicated still further by an engineer/manager/scientist/regulator/politician/activist tangle. One example of the kind of organizational tugs-of-war that continued throughout the crisis was control of the website dedicated to the oil-spill response. Whereas BP had overseen the deepwaterhorizonresponse.com site for the first two-and-a-half months after the explosion, the Department of Homeland Security, which oversees the Coast Guard, chose to change the site to a dot-gov one.[20]

At one point in the drama, Secretary Chu reportedly had sent to Houston a team of senior officials from national laboratories to work on figuring out why the blowout preventer failed to function properly. A group of "the world's top scientists" continued to work with BP in considering further options. While scientists eventually did use some advanced gamma-ray techniques to peer into the preventer, showing that a critical component known as a shear ram had only partly deployed, the scientists did not come up with any better solutions than the engineers. It was indeed a sticky problem. Early on, attention had become focused on the failure of the blowout preventer as the immediate cause of the problem and the last hope, short of a relief well, for a solution to stop the flow. Even the possibility that a relief well might fail to work would eventually come to be discussed publicly, with some insiders admitting the possibility that the leaking well was so badly damaged that the bottom-up scheme would be ineffective.[21]

In the meantime, theories about the mechanical cause of the well failure were being developed in parallel with the efforts to deal with it. The jammed "blind shear ram" had been designed to be the last line of defense in case of an emergency. But discovering that it was somehow jammed was scientific knowledge, not an engineering fix, nor even a useful lesson learned. The idea behind this recalcitrant component of the blowout preventer was that, should all other means of shutting off the flow of oil fail, the blind shear ram would cut through the drill pipe and seal the well completely. In the case of *Deepwater Horizon*, when the shear ram only partly deployed, it was rendered ineffective. An early theory for the failure was that, by chance, a joint in the drill pipe was located at just the same level as the shear ram blades. Since the joint consisted of a collar connecting two sections of pipe, an extra thickness of steel kept the shear ram from doing its job. Another theory, prompted by the gamma-ray images of the innards of the blowout preventer, was that somehow two separate sections of drill pipe had become jammed between the halves of the shear ram, thereby blocking its effective deployment. The purported second pipe may have been projected into the shear ram during the well blowout.[22]

A study of blowout-preventer reliability in deep water had been commissioned the year before the Gulf accident. Focusing on about 15,000 wells that had been drilled in the North Sea and off North America over about a twenty-five-year period, the study found eleven cases in which a blowout preventer was activated when the crew on a rig had lost control of its well. In five of the cases, the preventers were not able to control the well, which means that the failure rate was 45 percent, a number that should have been considered unacceptably high. (A burst hydraulic line and a shear ram failure were blamed for an inability to control the 1979 blowout of the exploratory oil well *Ixtoc I*, located off Mexico's Yucatan Peninsula, resulting in one of the world's largest spills then to date. Shear ram failures were also blamed for less consequential blowouts off the coasts of Texas and Louisiana, in 1990 and 1997, respectively. On the other hand, just four months before the *Deepwater Horizon* incident, a blowout preventer worked as it was designed to do when a Transocean rig operating in the North Sea experienced a pressure surge that led to oil and mud being released onto the rig. Activating the blowout preventer in this emergency did seal the well. This is how percentages and probabilities of failure manifest themselves.) Blowout preventers were not only unreliable but also expensive to maintain. The cost of stopping drilling operations to retrieve a preventer to repair it had been estimated to be about $700 per minute. When operated properly, shear rams were also to be tested regularly. A study of government-mandated testing of blowout preventers found only sixty-two failures in almost 90,000 tests, or a rate of less than 0.07 percent. While the effectiveness of the tests had been called into question—something that is supported by the failure rate under not test but emergency conditions—the oil industry had used such data to argue for reducing required testing frequency from fourteen days to thirty-five. It was estimated that such a relaxed testing schedule could save about $193 million per year. Such competing goals of cost and benefit complicate debates and decisions in industries of all kinds.[23]

Although the government, news media, and citizens focused on BP when assigning blame and responsibility for the Gulf oil spill, other companies were also called to task. In particular, Transocean, the owner of the *Deepwater Horizon* rig that had been rented to BP for $550,000 per day, came in for special scrutiny. More than three months into the spill, it came to light that Transocean had had, because of "a series of serious accidents and near-hits," its own concerns not only about *Deepwater Horizon* but also about the safety of some of its other rigs. Just a month before the fatal accident, the company had commissioned from Lloyd's Register a wide-ranging review of the "safety culture" of Transocean's North American operations. In particular, it wanted the risk-management firm to investigate the rig owner's headquarters in Houston and some of its specific oil rigs and provide confidential reports. When these reports were made public by the *New York Times*, it became clear that there had been preexisting problems that were not fully addressed. According to the reports, there existed "a stifling bureaucracy imposed by onshore management," which led to "widespread resentment among rig workers." This in turn led to "a significant mistrust between the rigs and the beach," the latter term referring to the management in Houston. The safety manual, it was claimed, "was written for the courtroom, not the oil field." Workers on the rigs were said to have feared reprisals if they reported problems, and major maintenance was thought to have been routinely postponed. Critical equipment could thus be operating in conditions close to failure. Some of the information in the confidential reports gave indications of what might have caused *Deepwater Horizon* to sink. Unaddressed issues with its ballast system, which was designed to maintain stability on the floating rig, may have led to the sinking that followed the explosion. Had the semisubmersible, even though damaged by fire, not sunk, the well pipe might not have ruptured, and the oil spill might not have occurred.[24]

But, of course, it did. And even as efforts to contain the oil leaking into the Gulf continued, congressional and other hearings, industry investigations, and preliminary failure analyses were taking place—resulting not in confidential reports but in daily news stories coming out of open meetings. One early conclusion was that, independent of the technology of blowout preventers, a lot of the blame for the spill could be traced to unwise decisions seemingly prompted by pressures to save time and money in the drilling operation. One BP employee described the *Macondo* project as a "nightmare well," which had given repeated warnings that more care should have been taken with it. A proposed scenario for the accident was that the well casing somehow developed a leak, allowing gas to rise up in the borehole. This situation went unnoticed by the crew on the drilling rig, which had misinterpreted pressure tests on the well proper to indicate that it was sealed. Believing this to be the case, the crew started displacing the heavy drilling mud that had been containing the well pressure with seawater. Soon, however, they noticed an unbalanced pressure and began to lose control of the well. Methane gas drove mud and water through the riser from the blowout preventer and shot the mixture several hundred feet into the air above the rig. The methane was ignited, perhaps by a spark from some equipment. When the blowout preventer was activated, it failed to do its job. The explosion was followed by the fire and the eventual sinking of the rig.[25]

One hearing before a government panel investigating the *Deepwater Horizon* accident revealed that the drilling rig's emergency alarm had been set to an "inhibited" mode, so that false alarms did not wake up sleeping crew members. This ill-advised action very likely slowed evacuation of the burning rig and may have cost some of the crew their lives. The panel also heard of failures—including leaks in emergency equipment, crashing computers, and losses of power—that occurred repeatedly in the weeks preceding the accident, further suggesting that the vessel was not well maintained. Indeed, it came to light that an audit conducted in 2009 had uncovered hundreds of needed repairs, as well as errors in judgment and operation. The documented history of mistakes and shortcuts had apparently continued even after the audit, for on the day before the accident an emergency equipment test was not conducted properly and so did not detect a gas burp that may have been a precursor to what caused the explosion. The separate audit conducted by a Lloyd's risk-management group just weeks before the accident found the "safety culture" of the rig wanting, with workers reluctant to report problems for fear of reprisals. Conditions on the *Deepwater Horizon* evidently had set the scene for an accident waiting to happen. It seemed to be just a matter of time.[26]

According to a review committee of the National Academy of Engineering, the "preventable" blowout was "caused by a complex and highly improbable chain of human errors coupled with several equipment failures." The chairman of the Senate Energy Committee attributed the explosion to a "cascade of errors" that included technical, human, and regulatory mistakes. A congressional subcommittee focused on a number of errors attributable to the rush to finish the job and thereby reduce costs by not following standard industry practice. According to testimony of industry experts and executives, these errors began with the design of the well itself. Rather than employing multiple well liners, BP used only one, which increased the chance that errant gas could reach the surface. The design also cut corners by using a smaller number of spacers to keep the drill pipe centered in the well bore. This might have prevented the casing cement from flowing uniformly into the annulus, allowing gas to escape through the unsealed space. (The presidential commission investigating the accident would find that as many as twenty-one "centralizers" had been recommended by a Halliburton engineer, and BP had intended to use at least sixteen. However, when it was determined that only six were available, they were all that were used.) Furthermore, when the cement was in place, it was not checked for leaks. Also, the wellhead was not secured with a lockdown sleeve, another standard industry practice. Such corner-cutting decisions may indeed have saved time and money, but they also increased the risk of a blowout. By taking these design shortcuts, one "veteran deep-water engineer" estimated, BP shortened the work by a week, amounting to a savings of maybe a million dollars a day, but the reliability of the well was significantly less than it could have been had the company taken a more conservative approach.[27]

From the beginning, environmentalists and Gulf Coast state and local officials stressed the ecological damage that the leaking oil could produce. They expressed repeated frustration both with absent and uncoordinated federal efforts, which were described as "chaotic," to keep the oil from reaching the shore and with cleanup operations, which could qualify as a collateral failure. An environmental scientist concerned about reports of deep underwater oil plumes, which he feared might not surface for years or decades in the future, characterized the situation as a novel one "riddled with unknowns." He called it a "scenario where reality is ahead of the science." But warnings about what might happen, whether to the natural environment or to a mechanical system, often go unheeded, especially when things appear to be under control. The Yucatan, Texas, and Louisiana incidents may have had lessons to share, but they were not widely remembered, perhaps because they had not broadly affected the U.S. Gulf Coast. In the absence of immediate bad experiences, it is human nature to assume that everything is going well and to relax oversight. But such behavior should not be regulatory nature. The Minerals Management Service, the bureau of the Department of the Interior that was responsible for regulating offshore drilling, came in for considerable criticism. The agency regularly underplayed environmental concerns and waived rules to grant quick approval of activities related to energy exploration. It also downplayed technical difficulties, barely mentioning blowout preventers in a report on the risks and benefits of increasing offshore drilling. One observer described the situation as "a war between the biologists and the engineers." After the accident, some government agencies understandably overcompensated, with the Environmental Protection Agency taking what some considered an inordinate amount of time to approve actions intended to keep the oil from reaching marshlands and other ecologically sensitive areas. Indeed, government agencies were apparently so overwhelmed with work in conjunction with the Gulf oil spill that evaluations of seemingly unrelated projects, such as a proposed bridge in Minnesota being challenged by the Sierra Club, were delayed.[28]

It had been three weeks into the spill when the public first got a good look at the oil spewing out of the broken pipe a mile under water, and this became the iconic video image of what was by then increasingly being termed a catastrophic event. It was difficult to judge the rate at which oil and gas were coming out of the uncontrollable source and mixing turbulently with the deep-sea water, but estimates had increased as the spill went on. Scientists debated the volume of oil being released; where it would be carried by winds, ocean currents, and hurricanes; and the predicted damage it posed to the ecology of the Gulf Coast. This interrelated knowledge and speculation lent a sense of urgency to the need to stop the leak and get the rate of discharge right. It was clear that it could take months or years to empty the reservoir of its estimated fifty million barrels of oil, and so knowing how fast it was emerging from the broken well was important for planning purposes, if for nothing else. Estimates of the flow rate from the well were based on analyses of videos showing the escaping hydrocarbons, computer models of the reservoir and well, and the amount of oil recovered by ships on the surface. The initial government figure of 1,000 barrels daily had been changed to 5,000 after about a week, then it was raised to between 12,000 to 19,000, the range emphasizing the scientific uncertainty of measuring the size of the leak. This new estimate made the oil spill the largest in U.S. history—larger than that leaked from the 1989 incident when the tanker *Exxon Valdez* struck a reef in Alaska's Prince William Sound. The effects of the *Deepwater Horizon* oil spill were also compared with those of Hurricane Katrina, which had devastated New Orleans and portions of the Gulf Coast in 2005. Still, the leak estimate was soon raised again, to 20,000 to 40,000 barrels. Eight weeks after the explosion, the estimate was revised upward yet again, to 35,000 to 60,000 barrels a day. The oil company would later challenge the government's official estimates, claiming that they were as much as 50 percent too high. Since BP was facing possible fines of $4,300 per barrel spilled, billions of dollars were at stake.[29]

Regardless of the exact amount of oil involved, while it was spewing from the well surface vessels tried to capture what oil they could. At the time, they did not appear to have enough capacity, and so the government asked BP to add more. However, even with the rising flow rate, the administration's claim that the spill was the "worst environmental disaster America has ever faced" was being challenged. Historians pointed out that the 1930s Dust Bowl was responsible for more social upheaval, the *Exxon Valdez* spill killed more wildlife, and the pesticide DDT affected a much larger portion of the country. Historians also noted the irony that the day the *Deepwater Horizon* sank was the fortieth anniversary of the first Earth Day, which was instituted at least in part to remind people to learn from their environmental mistakes.[30]

The Gulf of Mexico has been the scene of plenty of environmental insults. In the middle of the twentieth century, its approved dumping zones accumulated bombs, chemical weapons, and other ordnance that have remained on the bottom. Since the mid-1960s, the area has seen over three hundred oil spills attributable to offshore drilling, accounting for a total of over half a million barrels of oil and "drilling-related substances." *Deepwater Horizon* itself was responsible for four of these smaller spills. Even if its catastrophic spill, whose final volume was put at almost five million barrels, about 800,000 of which were captured by containment efforts, was not historically the number-one environmental disaster, it devastated the already fragile Gulf Coast economy, threatened to bankrupt a global corporation, and embarrassed the government —emphasizing the range of impacts that a technological failure can have.[31]

A spectrum of people, organizations, and design decisions came in for early blame. The congressional committee and other government groups, which began meeting about three weeks after the explosion, heard testimony from executives of BP, Transocean, and Halliburton, the contractor hired to perform services on the rig, including sealing the well with cement. Although Halliburton did perform the pre-accident cementing job—the procedure that was supposed to have sealed the well until BP was ready to start production operations—the contractor maintained that it had worked from BP's design. So the accusation was not only against the design but also against the design's designer. Regardless of whose responsibility it was, an ineffective cementing job was identified early on as a likely factor in causing the accident. Several possible problems with the cement job—carried out in conjunction with the use of a well casing that made the "best economic sense" but was described as a riskier option—were speculated upon: the cement provided an improper seal, which allowed gas to escape up the drill hole; it did not set quickly enough, again allowing gas to escape; it was a type of nitrogen-containing cement that was tricky to handle and so might not have formed a solid seal. (There were, in fact, reports of gas leaking through the cement in the hours before the explosion.) The last of these hypotheses was a focus of the House Energy and Commerce Committee and other groups. A definitive judgment on any hypothesis was naturally difficult to make without some of the evidence that lay beneath the sea. The 325-ton blowout preventer could be retrieved for forensic analysis once the well was capped; raising the *Deepwater Horizon* rig itself would be a tougher challenge, since it is so large. A conclusive testing of the most likely cement hypotheses might have to wait for recovery of the drill pipe, which is not likely ever to occur. In the meantime the hypothesis was inconclusively explored at the hearings.[32]

Alongside congressional hearings, an inquiry into the accident was conducted by a joint investigative panel comprising members of the Coast Guard and the Minerals Management Service, that part of the Department of the Interior that had been responsible for protecting the public and the environment while at the same time leasing and collecting revenue for the use of government-controlled oil fields. This was a clear conflict of interest, in which

mineral agency employees were rewarded for expediting the grant-
ing of licenses. In the case of the *Deepwater Horizon*, the Minerals
Management Service approved BP's application to operate the rig
in the Gulf even though the paperwork was missing a permit to
drill in the vicinity of endangered species or marine mammals.
Staff biologists and engineers were reported to have been "regularly pressured by agency officials to change the findings of their
internal studies if they predicted that an accident was likely to occur or if wildlife might be harmed." This kind of experience no
doubt affected the culture in which the staff worked. Also missing
from the *Deepwater Horizon* application was information on the
strength of the shear ram in the blowout preventer, which was
supposed to establish its effectiveness in the event of an emergency. The engineer responsible for reviewing the application admitted that he did not regularly check for that information—he
just assumed that the applicant was complying with regulations.
The petroleum companies were thus in effect allowed to certify
their own equipment.[33]

Problems with offshore oil drilling clearly went well beyond design. In conjunction with the inquiry the administration in Washington announced that the Minerals Management Service would
be split in two, with one office now responsible for public safety
and environmental enforcement and the other for the leasing and
revenue-collection functions. This decision was reminiscent of the
1975 split of the Atomic Energy Commission (AEC) into the Nuclear Regulatory Commission and the Energy Research and Development Administration, which soon thereafter was integrated into
the Department of Energy. While that split was not prompted by a
physical failure, the breakup of the Minerals Management Service
clearly was, although the example of the AEC should have been
enough of a lesson learned to divide the agency earlier.[34]

In an interim report of its preliminary findings, a committee of
the National Academy of Engineering and the National Research

Council charged with looking into events leading up to the explosion and subsequent oil leak identified as an underlying cause "a failure to learn from previous 'near misses.'" This cultural mindset was eerily reminiscent of that which had plagued the space shuttle program. In the case of the Gulf accident, the committee found that there were "insufficient checks and balances for critical decisions impacting the schedule" for sealing the exploratory well until it could be reopened for a production phase. There was no evidence of a "suitable approach for managing the inherent risks, uncertainties and dangers associated with deepwater drilling operations"; nor was there a "systems approach," whereby the various factors affecting the safety of the well and its operation could be integrated. The problem was, in short, that isolated decisions could be and were made without regard to their consequences for the ultimate safety of the whole. In its final report, the committee was expected to recommend ways "to establish practices and standards to foster a culture of safety and methods to ensure that schedule and cost decisions do not compromise safety." But long before that was to happen, other investigative bodies were to weigh in on the accident.[35]

A month after the *Deepwater Horizon* explosion, a bipartisan national commission had been named to look into the causes of the accident and the environmental destruction that resulted. One stated purpose of understanding how and why the accident occurred was to "make sure it never happens again." This was a customary and noble objective, of course, but it was also an unrealistic one. The only way to ensure that another offshore drilling accident—or any kind of failure—never again occurs is to cease exploration and innovation. In the meantime, the administration had imposed a six-month moratorium on deep-water drilling in the Gulf, but a federal court imposed an injunction on that suspension. Those opposed to the moratorium pointed to the severe economic hardship that the region would suffer as a result. Appeal followed appeal, and eventually the moratorium was lifted a month early but soon reinstated as an indefinite ban on expanding drilling into new areas. That reinstatement was then modified to allow limited drilling to resume in certain areas. Throughout all the back-and-forth on policy, what was generally absent from the public debate, at least, was a quantification of the risks involved.[36]

All technology and technological endeavors carry risk, and the experience with the *Deepwater Horizon* rig and its associated undersea equipment and operations and the management thereof not only highlights this fact but also shows how risk can be pushed to what should be unacceptable limits. When those limits are exceeded, failure eventually occurs, and unraveling exactly which component of risk contributed exactly what to the ensemble is no easy task. The full and true explanation of an accident like that which will long be referred to simply as the *Deepwater Horizon* or Gulf oil spill may never be known with absolute certainty. But even if past failures are completely understood, it is unlikely that future failures will be fully avoided. Not only will the future differ from the past and so contain the potential for new things to go wrong, but also human beings and their creations are likely to remain as imperfect as they have always been. According to the presidential commission, the accident was a "failure of management." Had the three companies involved in the drilling and well-sealing processes engaged in "better decision-making and risk assessment," the accident might never have occurred. The accident was ultimately attributed to "systemic failures," and, in the absence of substantial reform in the industry and its federal oversight, a similar accident could certainly happen in the future.[37]

The true problem with the *Deepwater Horizon* blowout preventer was attributed to a design flaw. When the well blew, the force of the oil stream evidently bent and knocked off-center the drill pipe, and this jammed the shearing mechanism. Consequently, when the blind shear rams were activated, their blades were immobilized short of coming together. They were left less than an inch and a half apart, but this was enough of a gap for the oil to gush through. The device had been designed to work effectively on a pipe centered in the well, even though the force of a blowout should certainly have been expected to be able to destroy that symmetry. In the wake of the accident the Bureau of Ocean Energy Management, Regulation, and Enforcement established more stringent safety measures, but those pertaining to testing blowout preventers did not require doing so under asymmetrical conditions. The extent of the potential failure modes that a regulator or designer has to anticipate will always be a matter of debate, revolving about questions of risk and chance and consequences. According to one student of the *Macondo* well blowout, it "happened because of many mistakes," and he believed that the faulty blowout preventer could be added "to the other 9, 10 or 11 reasons why it happened."[38]

This sentiment was reinforced by the final report of the joint task force of the Coast Guard and the Bureau of Ocean Energy Management, Regulation, and Enforcement. It concluded that the fatal explosion, prolonged oil leak, and consequent environmental pollution "were the result of poor risk management, last-minute changes to plans, failure to observe and respond to critical indicators, inadequate well control response and insufficient emergency bridge response training by companies and individuals responsible for drilling at the Macondo well and for operation of the Deepwater Horizon." Ultimate responsibility for the accident lay with BP, according to the report, but it did not absolve the contractors involved. A statement issued by BP expressed agreement with the report's "core conclusion," which it took to be that "the Deepwater Horizon accident was the result of multiple causes, involving multiple parties, including Transocean and Halliburton." The language signaled continuing litigation as another result of the blowout.[39]

As accidents like *Deepwater Horizon* recede into the past, lessons learned from them tend to be forgotten. Even though the names of the events may remain in the public memory, their nature and causes become as ignorable as the many weapons that lie out of sight on the bottom of the Gulf of Mexico, where they were dumped half a century ago. Just as the lessons of the space shuttle *Challenger* explosion were evidently not sufficiently learned to prevent the loss of *Columbia,* so it would be naïve to think that the aftermath of the explosion on *Deepwater Horizon* would be any more efficacious in preventing all future such occurrences. An oil executive-turned-regulator explained the deteriorating safety record of the offshore industry with the observation that "people can forget to be afraid." In other words, they can cease to suffer from kakorraphiaphobia, or fear of failure. But whether or not failure is feared, we can mitigate its effects by developing out of past failure experiences new devices, systems, and procedures that an enlightened technological culture insists be taken seriously. Unfortunately, when complicated technology and complex professions and flawed individuals interact we can expect failure to follow, maybe not today, or next week, or next year—but invariably it will happen. Indeed, the longer it does not, the more likely the human component of the machines/people system will believe that it will never happen again. That attitude breeds complacency, which leads to a relaxation of guards of all kinds—technological, political, organizational, and psychological—against failure.[40]

THIRTEEN

Without a Leg to Stand On

Children love trains; they also love cranes. When I was a child, one of my favorite model railroad cars was one that had a crane mounted on it. The crane's body swiveled, and the two cranks on its side raised and lowered the boom and the hook that hung from it. Derailments, which did not have to be simulated, were frequent, and the crane car was often brought into service to right the toppled rolling stock and lift it back onto the tracks. Cranes were also among my favorite Erector set building projects. My beginner set was limited in parts and accessories, and so the earliest cranes I made were limited in their size and capacity. As I dreamed of graduating to larger and fancier Erector sets, the cranes I constructed in my mind were taller, more mobile, and motor-driven. But children also grow tired of assembling and disassembling the same basic steel girders and pulleys and cranks in the warmth and safety of their home, and so they look outward for greater adventure—even if it means risking failure. Children learn a lot from play.

I spent my first twelve years of life in a section of Brooklyn where the houses were attached and the sidewalks were perforated. The subway grates were a source of great attraction for young boys, for out of them came the noise of approaching trains and the extended blasts of air that they pushed piston-like ahead of them through the dark tunnels. These were the updrafts that billowed skirts and dresses—including Marilyn Monroe's in a famous photograph—but for prepubescent boys these occurrences were less interesting than the air currents that lifted handkerchief parachutes tied to metal washers against the force of gravity. An even more lasting attraction to the subway grates, however, was the treasure that people accidentally dropped though them while rushing to catch the train heralded by the noise and the wind.

Depending on the station configuration, the floor of the plenum beneath the sidewalk grates could be fifteen or twenty feet below street level. Especially around midday, when the sun shone deep into the space, we boys could see the prizes that lay below: coins, keys, pens, rings, earrings. It became a pastime of childhood patience for us to fish for these valuables. The only equipment we needed was a sufficiently long piece of string, a weight—such as a small padlock whose key had long been lost—to tie onto the end of the string, and a piece of sticky chewing gum. (Sometimes a small magnet replaced the lock and gum, but it proved ineffective for retrieving most treasures.) Like a gantry crane that moves above a factory floor, we positioned ourselves over the object of our desire and lowered our line through the grid cell that we judged to be plumb above our target. This was a crucial step, for the one-by-two-inch cells allowed only limited horizontal fine-tuning. If we missed our mark by too much, we had to haul up the line and drop it through another cell.

We did get pretty good at spotting ourselves directly over our pelf, but that was only the beginning of the task. When the sticky gum looked like it was just above what we hoped to bring up, we would let the weight free fall for the last few inches to establish a bond. Then came the retrieval stage, in which the string was hauled up at a steady pace. However, like any system, this one was subject to failure at all stages of the operation. Indeed, as the prize—let us say it was a coin, as it usually was—approached the underside of the grate, several things could go wrong: the gum, which had become contaminated with dirt from near misses, might no longer hold the coin, and it would drop back to the bottom before it reached the top; or the coin, not being fully centered below the weight, might have an edge projecting out just far enough to be caught by the side of the grid cell and so be pried off and lost to gravity; or we might get the coin safely through the cell but in our eagerness to possess it find it slipping through our fingers and falling back down into the plenum. Especially if our target was a quarter, which in the horizontal position would barely fit through the grate anyway, we would watch its trajectory intently and move our crane body over its new resting place and try again. We learned from our failures, of course, and our success rate grew accordingly.

As hard as the steel grid was on our elbows and knees, we could spend hours on the subway grate, instinctively getting to know the frequency of the trains and their precursor air blasts that could also disturb our lines and dislodge our cargo. Our take for an extended stay might be a handful of change—keys and jewelry held little interest for us—but it seldom amounted to more than a dollar or so. But perhaps even more than the coins, which would be squandered on penny candy, the pleasure of the pursuit—and the enjoyment of the accomplishment—held our fascination and captured our imagination. We hovered like helicopters over a battlefield or sinking ship, hoisting up the bodies of the injured. We were human lifting cranes, skyhooks of a sort. Little did we know then that some of us would become engineers, responsible not for operating but for designing the booms and cables and controls of real construction cranes that were not supposed to fail in action.

Design is a desk job behind the lines; construction is combat on the front. The development of the plans for a long-span bridge or a high-rise building may proceed in virtual cyberspace accessed through a computer terminal in an air-conditioned office, but the realization of the completed design across a wide river or in the heart of a crowded city necessarily takes place in the real world, where risks are high and slips can be fatal. In cyberspace, which is devoid of mass and where information moves (theoretically) at the speed of light, weighty words and mind-numbing numbers can be cut and pasted from here to there in the wink of a mouse's eye. In Earth-space, however, where gravity rules and overcoming it takes time and energy, tens of thousands of miles of steel wire may have to be played out between anchorages and over skyscraper-tall suspension-bridge towers, and thousands of tons of steel girders may have to be lifted into the air and fitted together with a feather-light touch and at a maddeningly deliberate pace above busy shipping lanes. Similarly for a tall building, heavy steel beams and columns must be lifted up from and moved over bustling streets of commerce.

Construction cranes have been used since ancient times. They were originally made of wood, and the earliest ones consisted of a pair of timbers lashed together at one end, where a pulley or block and tackle was also located. A guyline held the timbers at the desired angle to the ground so that the top could be positioned over whatever was to be lifted or repositioned over where it was to be set down. At its base, the pair of timber legs was spread apart to provide some sideways stability. The rope by which a lifting force was exerted was wound around an axle located between the legs, and it was turned by means of a crank, capstan, or treadmill. Relatively small cranes could be operated manually by the crank, but larger ones required the treadmill, in effect a large squirrel cage in which not mice but men climbed an endless ladder. Basic crane types were described by Vitruvius, and the timbers used as described have come to be known as shear legs. Wooden cranes naturally evolved into more elaborate forms with greater capacities, and they were used in cathedral construction during the Middle Ages and even into modern times.[1]

The concept of lifting devices was widely illustrated in Renaissance notebooks, treatises, catalogs, and so-called theaters of machines. Leonardo sketched what we can recognize as cranes; Agricola's *De re metallica*, first published in 1556, is full of illustrations of devices for lifting ore and other materials out of mines; and Agostino Ramelli's 1588 book of diverse and curious machines has its share of lifting devices. But as late as the Victorian era, simple shear legs were still commonly used. With the development of cast iron and its use in bridges and later buildings, large structural arch ribs, columns, and beams could be erected with shear legs and block and tackle and were relatively easy to hold in place until the rest of the structural components could be raised and everything joined together to make a strong and sturdy self-supporting assemblage. This is evidently the way the first iron bridge, understandably known simply as Iron Bridge, was erected in 1779 near Coalbrookdale, England, and how the wrought- and cast-iron girders of the Crystal Palace were erected in London's Hyde Park for the Great Exhibition of 1851. Wrought-iron cranes and derricks were also developed for use in loading and unloading ships tied up at a dock, and steel ones came into use in the late nineteenth and early twentieth century. Mobile and powered cranes fitted with buckets rather than hooks, and known as steam shovels, were also soon operating and would prove to be invaluable in the construction of the Panama Canal.[2]

Cranes are also useful teaching tools. Homework problems involving derricks, cranes, and their taller relatives—broadcast towers and guyed masts—are commonly encountered in an introductory engineering course, because they can be at the same time conceptually simple and geometrically challenging. The three-dimensional nature of the problems can test the novice's spatial skills and trigonometric talents. Even though the principal function of a crane is to move heavy objects from one place to another—usually higher or lower—the motion involved is supposed to be relatively slow and deliberate. This not only makes for safe operation and more accurate control but also eliminates the potentially uncontrollable and destabilizing effects of inertia associated with sudden movements. In engineering terms, the acceleration and its accompanying forces are so small as to be ignorable. Thus, according to Newton's second law of motion, which states that the net force on a body is proportional to its mass times its acceleration, the forces are effectively all in equilibrium with one another.

A modern crane is essentially a mechanically operated winch or windlass in combination with a projecting arm known as a boom or a jib, the base end of which usually pivots and rotates, and the other end of which is most basically fitted with a pulley or sheave over which a cable passes, just as a line did in a pair of shear legs. This enables the force supporting a load to be redirected from the vertical defined by gravity to the angle of the boom directing the hoisting cable back to the winch. In recent times, construction cranes have grown in complexity, size, capacity, and versatility. The largest of them were in the news often in 2008, principally because that year they were involved in a number of highly visible and deadly accidents. Construction work of all kinds is a dangerous activity, with an average of about four worker deaths occurring each day across the United States. The total number of construction fatalities did fall in 2008 and 2009, but this was attributed more to unfavorable economic conditions, which resulted in fewer hours worked on construction sites, than to any other factor. In crane accidents alone, in a recent year, as many as eighty-two workers were killed, and many more injured, with most of the incidents getting little notice beyond the local news and trade press. On average, almost ninety people die annually as a result of "crane accidents and related hazards." But several of the crane accidents that occurred in 2008 received unusually broad news coverage because they reached beyond the construction site. As much risk as working in the construction industry may involve, that risk is not expected to extend to ordinary citizens going about their daily activities on the streets below and in the buildings beneath or beside the looming cranes.[3]

The large mobile crane, capable of moving under its own power on wheels or caterpillar crawler tracks to almost any location on a construction site, is essentially a self-propelled heavy-lifting device, albeit one that may have to extend outriggers and counterweights to balance itself securely for wide-ranging work. But such cranes are limited in how far their boom can reach and thus how high and wide they can hoist. To construct buildings taller than the tallest mobile crane, some kind of hoisting system obviously has to tower over the incomplete structure or rise with it. In the case of the Twin Towers of New York's World Trade Center, whose construction began in the late 1960s, four cranes were mounted on the growing steel framework of each building, and these were repositioned to a higher floor as the structures rose. The technology originated in Australia, and the hopping devices were known as kangaroo cranes. Each of them had a jib capable of servicing one of the overlapping quarters of the floor area of the square tower on which it was mounted.

The use of such ingenious bootstrapping systems requires careful planning, not only to raise the cranes as the building itself rises but also to return the cranes to the ground when the building is done. In the case of the World Trade Center, one of the four cranes could be used to lower the parts of the other three to the ground, but naturally the last crane standing could not easily lower itself. It was thus essential that it could be lowered by a smaller crane that could be disassembled into parts individually small enough to fit into a service elevator. Cranes similar in design to the kangaroos were employed in the United Arab Emirates in erecting the tallest building in the world, known as Burj Dubai while under construction but renamed Burj Khalifa at its dedication in 2010. The logistics of moving cranes up that building as it rose—and as it changed from a concrete to a steel structure at its 156th floor—was something that its structural designers clearly had to take into account. And it was only the most conspicuous of the city's construction projects dependent upon nimble cranes. Booming Dubai, which until its economy along with that of most of the rest of the world became depressed, had been called "perhaps the world's best-known design play land," in no small part because of the presence of a multitude of towering construction cranes.[4]

Throughout the last two centuries especially, with the accelerated development of larger and larger structures, including new forms of long-span bridges and tall buildings, design engineers have increasingly had to consider how the unprecedented constructions would be erected. In the early part of the nineteenth century, suspension bridges often were designed to employ chains of wrought-iron links, which could be lifted into place one at a time and joined upon scaffolding. John Roebling devised a means of spinning thousands of miles of individual steel wires into massive suspension cables, which by their length and size would be too heavy to lift into place if fully formed on the ground. He employed his scheme for his later bridges, and the method of construction was an integral part of the design. Obviously, if the construction method fails, so does the entire project. The essential features of the Roebling method of spinning cables continued to be employed by the John A. Roebling's Sons Company in the construction of many landmark long-span suspension bridges, including the George Washington and the Golden Gate.[5]

In the twentieth century, concrete bridges became popular, and by the end of the century long spans were being designed and constructed. Especially where the bridges were high above a valley or over wide bodies of water, builders increasingly needed to come up with innovative construction methods and the associated crane-like equipment for use in the erection process. In some cases, the equipment was a large structure in its own right, sufficiently strong and stiff to lift and transport massive precast-concrete sections of roadway out over the void before they were let down upon two adjacent piers. In other schemes, often employing large barge-mounted cranes, shorter sections of precast bridge segments would be added alternately to one side and the other of a pier, always in a balanced fashion and in some cases cantilevering out hundreds of feet before they were joined with another cantilever reaching out from the next pier. Designing such elaborate construction schemes to be fail-safe is clearly as important as designing the structure itself, and specialized cranes are essential to the process.

Today, the most visually striking type of crane employed in large-bridge and tall-building construction is the tower crane, which consists of a vertical mast that often rises up from the interior of or right beside the structure it serves and is topped most strikingly by a rotatable horizontal boom along which a hoisting hook travels to access any point within the circle of its reach. The first tower cranes were made in the early twentieth century; the first self-climbing ones date from mid-century.[6]

Tower cranes with horizontal booms are also known as "hammerheads." In the evolved version of these, which have their origins in shipyards and docks in the early twentieth century, the crane operator sits in a cockpit near the top of the mast and between the diametrically divergent boom and its counterweight-laden counterpart, and he may or may not have a 360-degree view of the activity below. However, with the aid of radio directions if necessary, an experienced operator can bring a load of materials within inches of any location on a rising building. These stationary (but not immobile) machine-structures have become familiar sights on large construction projects around the world. There are also tower cranes known as "luffers," which have a boom whose inclination to the mast can be varied. From a distance, these look more like a ground-based crawler crane nesting atop a fragile pedestal. Luffers are much better suited than hammerheads to performing hoisting work in confined construction sites, as are often found in densely developed cities like New York. Whatever form they take, in the late twentieth and early twenty-first centuries the number of construction cranes that towered over a city came to be seen as a measure of its economic health and industrial development. It was especially the T-shaped, lattice-framed hammerhead structures, which seemed to balance on one skeletal leg and appeared to pirouette on a hidden toe (and looked to grow as if lifting themselves up by their own bootstraps) that had come to be so pervasive that they raised many an eyebrow about how they worked—and how they all too frequently failed.[7]

And there were plenty of cranes potentially poised to fail, and not just structurally. At one time, a good portion of all the construction cranes in the world were believed to be operating in the United Arab Emirates. In 2006, it was claimed locally that Dubai alone had as many as 2,000 tower cranes in use. Boosters claimed that as many as 25 percent of the world's population of approximately 125,000 construction cranes of all kinds were operating there, but suppliers of the cranes put the figure much lower. One tourist called that city the "Land of Cranes," with the hammerhead types visually dominating the skyline, as they are wont to do. Their presence is so powerful a symbol of prosperity and growth that their population is commonly and deliberately exaggerated. Nevertheless, the abundance of construction cranes at work in Dubai, coupled with lax government regulation, poor maintenance, and communication problems among a multilingual and multicultural imported workforce, made accidents common but their reporting uncommon.[8]

The mechanics of a hammerhead tower crane are relatively

In 1999, this cluster of construction cranes was serving the building site of the Three Gorges Dam across China's Yangtze River. The cranes sitting on vertical masts, whose height can be adjusted as a construction project proceeds, are susceptible to failure during the dangerous process of raising or lowering them. In 2008, a rash of construction-crane accidents in the United States brought renewed scrutiny of the procedures by which the machines were regulated, inspected, maintained, erected, operated, and dismantled.

straightforward. The vertical mast section is typically anchored to a large concrete footing, which provides a firm foundation that can bear the weight of the crane and anything it might lift. The mast serves the obvious purpose of elevating the horizontal boom of the crane to a height sufficient to clear whatever is being constructed beneath. The horizontal element consists of two basic parts: the main boom, from which loads are hoisted, and the counterweight boom, which serves as a counterbalance, thus keeping the crane steady on its footing. The connected booms rotate together about the mast, so that they are always in alignment. When necessary, cables or tie bars attached to the top of the tower help keep especially long booms straight and horizontal. In order to lift and place a load anywhere on the construction site below the crane, the boom assembly must rotate about the vertical axis through the mast, and the lifting fixtures and counterweights must be mounted on trolleys that move in and out along the length of the boom.

The amount of material that a tower crane can lift depends on how far from the tower pivot the load is located. The farther it is from the supporting mast, the greater the tendency of a load to overturn the crane. This means that the capacity of a tower crane is not an absolute but is rather relative to the position along the boom of the hauling cable, measured from the vertical axis of the mast. In mechanics, the product of a force and a distance is known as the "moment" of the force. "Moment" in this context refers not to an instant in time but rather to the tendency that the force has to bend, twist, rotate, or topple a structure. (At the center point on the mast, the moments of the load being raised and of the counterweights ideally cancel each other; hence there is no net tendency for the mast to tilt.) If a tower crane were to attempt to pick up something many times heavier than its capacity, the moment of the force exerted by that load on the hoisting cable could cause the boom arm to buckle. This in turn would destroy the balance between the load boom and the counterbalance boom and would throw the whole crane out of equilibrium, causing it to topple over. (To avoid such an occurrence, a modern tower crane is fitted with an elaborate system of computerized controls incorporating limit switches.) The lifting capacity of a tower crane is specified in terms of the moment its main boom can sustain. Thus a crane with a sixty-meter main boom that can lift fifty metric tons at that distance would be designated as a 3,000 tonne-meter crane. The closer the load is to the mast, the heavier it can be. Because most large cranes are made outside the United States, the metric system of measurement predominates in their specification.

At the turn of the twenty-first century, among the largest tower cranes in the world were those manufactured by Krøll Cranes, a firm based in Denmark. The record holders were designated Krøll

K-10000 tower cranes, and a standard K-10000 could lift as much as 120 metric tons at 82 meters from the center of the crane (120 × 82 = 9840, or approximately 10,000 tonne-meters, hence the name). A long boom version of the K-10000 could lift up to 94 metric tons at 100 meters, which in familiar American units would be 104 so-called short tons at 330 feet. In other words, the lifting boom was longer than a football field. These cranes were made in limited quantities; as of the end of 2004, they had numbered only fifteen worldwide. In the meantime, Krøll had designed a larger crane, the K-25000, which carried a price tag of $20 million, but as of late 2011 not one had yet been purchased. In order to keep forces on such behemoths essentially within the realm of equilibrium mechanics, which is known to engineers as statics, a crane like the K-10000 swings about its tower at such a slow rate that it takes two and a half minutes to make a complete revolution. This not only eliminates any acceleration forces but also allows plenty of time for the load to be hoisted up, an operation that can proceed at a rate of less than twenty feet per minute for a maximum load. A light load can be raised at speeds almost ten times as fast, but it would still take at least several minutes to hoist something to the top of a very tall building under construction.[9]

There is not much advantage—but considerable risk and expense—in having a crane taller than the immediate needs of a construction site, and so tower cranes typically grow in height as the building grows beneath them. The initial assembly of a tower crane for a building site is commonly done with the assistance of a mobile ground-based crane, which necessarily limits the height of the mast as initially installed. As the building rises beneath the tower crane, there naturally comes a point at which the mast itself must grow in height for the crane to remain effective. Workers accomplish this "climbing" (or "jumping") of a tower crane by employing hydraulic jacks to open up a perhaps twenty-foot gap in the mast and then inserting a new mast section into the space.

With the new section in place, the hydraulic jacks are then used to lower the tower top down onto the new section, at which time it can be secured to the rest of the mast. (Very large towers, like the K-10000, may be topped by their own service crane, which can be used to assist in such operations.) Because of the automatic limit switches and other safety devices that prevent a tower crane from being overloaded during normal use, the most dangerous operations associated with a tower crane tend to be those during which the crane is being made taller or shorter.

When the economy was strong, as many as 3,000 tower cranes were operating in the United States at one time, making over 105,000 lifts daily. Like so many aspects of ubiquitous technology, construction cranes tend to be ignorable parts of the built and building environment—until an incident in which something goes terribly wrong. Perhaps not surprisingly, the vast majority of crane accidents are the result not of mechanical design but of human error or human foolishness. In one notable case, which occurred during the summer of 2005, a large crawler crane was being used on a condominium project in Jacksonville Beach, Florida. The crane had a main boom in excess of 100 feet long and, fitted to its top end, a luffing jib, which extended the reach to 140 feet. On the day of the accident, the crane was manned by a new operator who was standing in for the regular operator after he injured himself off the job.[10]

As is typical, the crane's boom and jib had been stowed in an off-configuration for the night, and so when the replacement operator started up the crane for the next day's work his first task was to reconfigure it for lifting. According to the crane's user manual, boom and jib were not to be operated at the same time. To help ensure that this rule was followed, the system computer was programmed to stop all motion if either jib or boom exceeded safe-angle limits. However, to allow the crane's components to be assembled, disassembled, and moved to or from a stowed position, the control panel was also fitted with a limit-bypass switch. To activate the bypass feature, the switch had to be held down with one hand while the other hand operated either the boom or the jib—but not both simultaneously. In the Jacksonville Beach case, however, the crane's regular operator had wedged a penny into the limit-bypass switch, thus keeping it thrown and freeing his two hands to operate both boom and jib at once. (Such clearly dangerous practices are neither unknown nor untried in the construction industry; sometimes a bypass switch is taped down to achieve a similar end.)[11]

Upon starting up the crane in the morning, the replacement operator did notice the penny wedged in the switch but left it there. According to a subsequent accident report from the Occupational Safety and Health Administration (OSHA), when he later noticed that the system computer was malfunctioning, the crane operator took the penny out, but then everything stopped working entirely. The balky system was reported to a sales representative of the crane company, and the operator was instructed to leave the penny out of the switch and reset the computer. With that done, the operator began to work the crane, but after a while the computer again began to misbehave. Instead of seeking more advice, he reinserted the penny into the limit-bypass switch, which allowed him to continue to operate the crane for the rest of the day. He put the crane into a rest configuration that evening. Upon returning to the job site the next morning, as he later related, the operator was pressured by his supervisor to make a few lifts with the balky crane, and it was being set up to do so when the accident occurred.[12]

In his apparent ignorance or disregard of the penny's role in circumventing one of the crane's safety features, the operator placed both hands on the boom and jib controls and operated them simultaneously, perhaps thinking that other limit switches would prevent him from achieving a dangerous configuration. Although a limit switch was to shut off power if the main boom exceeded an angle of eighty-eight degrees, with the bypass switch wedged in the on position the boom went beyond that and, as it reached the vertical, the overstrained cables snapped, the boom buckled, and the jib fell onto the building under construction, injuring several workers. Insurers have estimated that such operator errors account for about 80 percent of crane accidents.[13]

This incident repeats a familiar scenario: continued operation of a technological system despite its misbehaving. The situation is aggravated by pressure from management to move forward, despite clear warning signs that to do so is to tempt fate. As we have seen, the fatal launch of the space shuttle *Challenger* went ahead in cold weather even though there were numerous indications that the O-ring seals were unreliable under such conditions. Two dozen successful missions before *Challenger*'s January 1986 fatal one had given NASA as an institution an overconfidence that space shuttles could fly safely even with the erratic behavior of heat-shield tiles, fuel pumps, and O-rings. When a redesigned shuttle fleet resumed flights after the *Challenger* accident, there remained the pesky problem of insulation separating from the external tank during takeoff. This persistent problem came to be accepted—until one of those pieces of foam struck one of the wings of *Columbia*, which disintegrated upon reentry as a consequence of the damage suffered. Human nature seems to make us optimistic about systems that are warning and begging us to be pessimistic.

Early in 2008, two widely reported tower-crane accidents occurred during the critical heightening operation of jumping, and this brought increased scrutiny to the way large cranes are supervised and regulated. On March 15, a Saturday, a luffer tower crane that had been servicing a construction site on the East Side of Manhattan was in the process of being jumped so that it could hoist materials higher than the twenty-two-story height at which it had been operating. During this operation, the crane's mast was

to be steadied laterally by being tied into the rising building structure, and this was to be done with the help of massive steel collars.
At the time of the critical jumping operation, bracing collars were
already in place at the third and ninth floors, and workers were in
the process of adding a third one at the eighteenth floor. According to accident reports, that collar was being hoisted into place
when it suddenly broke free. The six-ton collar slid down the crane
mast and knocked the ninth-floor collar loose. The two falling collars then struck the third-floor collar, dislodging it also, and all
three came to rest near street level. With its bracings removed,
the crane mast was unstable and it fell across the street, crashing
through an apartment building on the other side. In all, seven
people were killed and two dozen injured.[14]

Within days, the root cause of the accident was identified as a
faulty yellow nylon strap that was being used to hold the top collar
up while it was being secured to the building frame. Lifting devices
of this kind are supposed to be checked for cuts, tears, and general
wear before each use, but such precautions are not always taken in
the rush of keeping to a schedule. Also, construction sites are often
full of dirt and mud, which can cake onto a lifting strap, not only
masking its true color but also any tell-tale imperfections. Whatever else may have happened in New York, photographs of the accident scene clearly showed a "yellow nylon sling ragged at the end
like a child's shoelace" that perhaps was being employed beyond
its useful life.[15]

Barely two weeks after the New York crane accident, another
occurred in Miami. This also involved a tower crane that was being used for the construction of a high-rise building. The crane's
height was being raised when a twenty-foot, seven-ton section being hoisted up for use in the climbing process fell about thirty stories and crashed through the roof of a nearby house that was being
used as a contractor's office. Two construction workers were killed
and five others injured in the mishap. Even before OSHA com-

pleted its investigation of the accident, it was being attributed to a failure in the rigging used to hoist the mast section.[16]

These two accidents, coming so close to each other and not long after other accidents in previous years, brought considerable attention to the dangers of tower-crane use, especially during the sensitive climbing process. Investigators of the accidents drew distinctions between a "crane accident" and a "rigging accident," the Miami and New York incidents falling into the latter category. Still, from a public perception and safety point of view, when the chains, slings, straps, and other paraphernalia—the rigging—that tie a load to a crane hook fail, the accident is understandably associated with the operation of the crane.

Only about ten weeks after the first incident, there was still another tower-crane accident in New York. No jumping operation was going on this time—the crane assembly and operator cab located above the turntable atop the mast just suddenly broke free and crashed onto the street below, damaging neighboring apartment buildings on the way down. The crane operator and another construction worker were killed. The accident was attributed to a broken connection at the turntable level, and a bad weld in a repaired steel part was immediately suspected. The incident fed growing concern in New York and elsewhere that tower cranes posed an unacceptable risk in large cities where construction went on high above sidewalks and streets crowded with unsuspecting pedestrians and unarmored vehicles.[17]

In the wake of these and other accidents involving construction cranes, there were calls for renewed scrutiny of operation and inspection procedures—and new laws and ordinances to govern them. An *Engineering News-Record* editorial, while acknowledging that it was "difficult to draw conclusions from random failures resulting from separate causes," did note that crane work was not rocket science and welcomed the new rules being introduced in cities where failures had occurred as being better than "the lax oversight looming over the rest of the country." In Texas, for example, where crane workers did not have to be certified, during the years 2005 and 2006 alone there were 26 crane-related deaths, compared with a total of 157 throughout the rest of the country.[18]

Elsewhere, at least, the problem of corruption seemed to be a contributing factor. Within days of the March New York accident, a city worker was arrested for falsifying an inspection report stating that he had visited the construction site in response to a complaint that the crane was not properly supported. The following week, the city proposed that in the future a city inspector would "have to be present every time a crane is erected, jumped or dismantled." The following month, the head of the city's buildings department, an architect by training, resigned after admitting that a permit should not have been issued for the building involved in the second crane collapse. In the course of seeking to fill the vacant position of buildings commissioner, the mayor's office suggested downgrading the requirements that the person be a registered architect or licensed engineer, so that the post could be filled by "the best person for the job, not just the best architect or engineer." Not surprisingly, the professional communities objected. And as surely as lawmakers attempted to control more closely the use and regulation of cranes, so could representatives of the construction industry be expected to oppose restrictions. A new safety ordinance in Florida's Miami-Dade County, in which hurricanes are not uncommon, required that a crane be able to withstand 146-mile-per-hour winds. According to crane owners, who rent the equipment to construction projects, the new standard demanded more "risky crane jumps and tie-ins to buildings," something that not only takes time and money but also endangers workers.[19]

As basic as the analysis of forces and moments on towering cranes might be, their safe use depends to a large extent on the willingness of operators to follow good practice and not to try to circumvent safety features like limit switches. But operators are also at the mercy of riggers, who can prove literally to be the weak link in the system if they use connecting devices that are worn out, damaged, or otherwise weakened by repeated use and misuse. A mud- or oil-covered sling that is too dirty to inspect properly before use obviously should not be used, at least not until it is cleaned up and inspected. And a rigger should know not to use a clean but faded synthetic sling that has been exposed to the sun over the years, for ultraviolet rays weaken the polymer material. The cost of a new yellow nylon sling would be pennies compared with the multimillion-dollar liability that could be incurred should the rigging break while lifting a climbing collar or hoisting a heavy load.[20]

New York's Department of Buildings ordered a year-long investigation into the rigging accident, and it found that one of the polyester slings that was being used in the jumping procedure was indeed damaged—it was frayed. Evidently that strap did not even have to be used, because, according to testimony, the day after the accident new ones were found in a tool shed. Furthermore, the rigging crew had used only four anchor points instead of the eight recommended by the manufacturer for lifting the collar. Since the manufacturer's specifications are effectively part of the jumping process design, and presumably have taken into account the forces involved in the procedure and what can go wrong, not following them is risky business. Failure to follow the rules in this case demonstrated "the consequences of taking shortcuts on the jobsite." Had eight slings (even including the frayed one) been properly anchored and employed as recommended, the factor of safety inherent in the process would likely have obviated the accident. The failure was especially tragic because less than two years earlier in New York a section of a tower crane had fallen to the street during a dismantling operation. The accident, in which three construction workers and two passersby were injured, was attributed to inappropriate rigging. Had the lessons of that failure been heeded, the fatal accident of 2008 might not have occurred, and manslaughter charges might not have been brought against the master rigger and owner of the rigging company. During the rigger's trial, the attorney for the defendant maintained that his client had been following "industry norms" but that "the crane was vulnerable because of engineering and design decisions" that were out of the rigger's control.[21]

The engineer responsible for the erection plan for the crane testified at the trial, and although he attributed the collapse to the failure of the slings used, he also spoke in support of the rigger, calling him "one of the safest riggers working." In the end, the rigger was acquitted of all charges by the judge, who at the request of the defendant had tried the case without a jury. His lawyers had successfully argued that their client had been made the scapegoat for a range of errors that contributed to the accident, including that the base of the crane had not been bolted to its foundation, that "shoddy" steel beams had been used to brace the crane's tower to the building it was servicing, and that there was lax and inadequate oversight by city inspectors. As is so often the case after an accident, many potential contributing factors could be identified and blamed for the failure. Perhaps no single one of them could be identified as *the* cause, and in fact the accident might not have happened if only one or two of the factors had been present. However, the coincidence of all of the contributing factors being present in this particular case created a perfect storm of individual failures that produced the colossal one that claimed seven lives.[22]

The other 2008 New York crane accident, in which the boom fell off the mast and onto the street below, was traced to a faulty weld repair of a cracked turntable part. Since the crane model was no longer being made, a new part could not easily be had. The cost of making a replacement part was estimated at $34,000 by one company and $120,000 by another, but their delivery time ranged from seven months to two years. The crane owner found a Chinese company that would do the job for $20,000 and deliver the part in three months. He went with the cheaper price and faster delivery, even though the Chinese firm itself admitted that it did not have "confidence on this welding." The owner of the New York company, a "prominent crane supplier" who was described as the "king of cranes," was charged with manslaughter. In the wake of the multiple crane accidents—the same crane company, which one observer said had "simply cut too many corners," had also owned the one in which the rigging failed—prosecutors were said to be pushing to hold owners of businesses personally accountable. Shortly afterward, the city inspector who had approved the crane with the faulty weld for use in New York was accused of neglect of duty and resigned.[23]

In a scandal unrelated to the failed cranes directly but publicized in the wake of them, New York's chief crane inspector admitted to taking bribes. He accepted money for completing and filing false inspection reports for cranes—though not tower cranes—that had only been looked over in a perfunctory manner or had not even been inspected at all. He also had certified that applicants for crane operator licenses had passed exams when in fact they had not. (Some crane operators were also indicted for possessing licenses without having taken a test at all.) These revelations led, understandably, to new regulations on the use of cranes at construction sites. The cranes and derricks division of the New York buildings department announced that henceforth detailed rigging plans would have to be provided by engineers or crane manufacturers before tower cranes could be jumped. The federal Occupational Safety and Health Administration also announced that it would investigate crane safety and propose operator certification. Once again, the occurrence of failures and deaths had led to measures to reduce future risk.[24]

The new OSHA crane-safety regulations, which replaced those that had not been updated for almost forty years, were issued in the summer of 2010. Among the most far-reaching changes contained in the thousand-page document was the requirement that construction-crane operators be certified on the equipment they operate. Since this rule affected an estimated 200,000 workers, companies were given until 2014 to comply with the regulation, which meant providing certification at their own expense. The new rules also took into account new technology and products, like synthetic slings, which had come into use in the meantime. With regard to slings, the new regulations required riggers to follow manufacturers' instructions in attaching them to loads to be lifted. In addition, tower-crane components were to be inspected before erection, and mandatory procedures were to be followed in operating cranes in the vicinity of power lines, where the largest number of lives had been lost in construction-crane accidents.[25]

Of course, risk can also be reduced by curtailing activity. The economic recession that became evident in the latter part of 2008 greatly affected construction, with numerous projects canceled, put on hold, or downsized. Since the typical crane was rented for a particular construction project, when that ended—because it was either completed or abandoned—the usual practice was to disassemble the crane and transport it to the next job. When there was no next job starting up, there obviously was no place for the crane to go. Around the world, tower cranes stood idle over incomplete structures, mute reminders of what might have been. Those locations that once boasted about the number of cranes that were silhouetted against the rising and setting sun now looked like crane parking lots. Cranes stood motionless above former boom cities like Dubai and Las Vegas. Whereas the large number of cranes that once pirouetted on their skylines gave them an ever-changing aspect that was invigorating to citizens, investors, and entrepreneurs alike, in 2009 the unchanging positions of the locked trusswork presented a tableau of stasis. The mobiles had become stabiles.

Thus it was also in the construction industry generally, where during early 2009 the unemployment rate stood at over 20 percent and was expected to go higher. Amid talk about stimulus spending and infrastructure investment, construction companies and their employees were looking for work. When public-works projects were advertised, contractors desperate for work were so competitive in their bids that they often came in below government estimates. A year later, construction-industry unemployment stood at about 25 percent and was feared to be still rising. The financial failures and bailouts associated with Wall Street and the pervasive sense of caution in the larger business community had combined to put a damper on development and so a stop to private-sector construction.[26]

Just as an undetected faulty weld in a steel turntable had led to a tower crane crashing down upon the street, so unrecognized flaws in the financial system led to a fiscal crisis. In 2010, Federal Reserve Bank chairman Ben Bernanke was reported to have admitted that "he had failed to recognize flaws in the financial system that amplified the housing downturn and led to an economic disaster." Bernanke said that he did not realize "the extent to which the system had flaws and weaknesses in it that were going to amplify the initial shock" of the collapse of subprime mortgages. The lessons to be learned from the fiscal follies were every bit as powerful as those to be learned from the crane accidents, but it remained to be seen whether either would be heeded when business and building returned to their pre-recession pace. One can only hope that after the period of inactivity lessons learned from crane accidents, from structural and mechanical failures in general, and from organizational, managerial, and financial dysfunction would not be forgotten. But experience tells us that, even though the slow economy allowed for plenty of time to think and reflect upon the causes of the devastating failures, human nature was sure to intervene and produce a lot of denial, if not downright amnesia, about what had transpired.[27]

FOURTEEN

History and Failure

Everyone, it would seem, should want to draw upon the lessons of the past in order to have a more successful future, but we do not always appreciate where the most valuable lessons are likely to be found. Shortly before President-elect Obama took the oath of office, the five living U.S. presidents—past, present, and then to be —assembled for some historic photos. As they stood before the press in the Oval Office, there was an understandable emphasis on success. Then-President Bush, speaking also for his predecessors, said they would be sharing their experiences with Mr. Obama, and that they all wanted him to succeed. Obama, who certainly wanted to succeed, said he hoped to learn from their successes.

Unfortunately, there are definite limitations and potential dangers in attacking the future with experiences of past successes alone—unless the desire is simply to replicate them. While the president-elect's comment may have been politic, he had run his campaign on the promise of change and already indicated that he wished to set a different course for the ship of state whose wheel he would soon take. But steering any ship through troubled waters can be a risky matter. One can succeed by following established routes, but even past successes may have depended upon luck as well as upon skill. What was a safe route through the past may not continue to be one into the future. Always, we must keep an eye out for trouble ahead—such as a looming iceberg.

In 1912, the innovatively designed ocean liner *Titanic* was heralded as an "unsinkable" success even before the ship was launched. As we all know, it sank on its maiden crossing. But now, a century later, let us engage in a thought experiment. Let us assume that the *Titanic* did not have the bad luck of being in the same place at the same time as a giant North Atlantic iceberg. Had the ship not had its unfortunate encounter, it might have reached New York safely and the success of its design would have been "proven." The more times the *Titanic* crossed and recrossed the ocean, the more confident the ship's captain, owner, and potential passengers would have become in its extraordinary seaworthiness. Competing steamship companies would likely have wanted to emulate the *Titanic*'s success, but they would also have wanted to make distinguishing changes that they believed would be improvements, whether for technical, economic, or commercial advantage. Larger, faster, and more opulent ocean liners would then likely have been designed and built. To make them more competitive financially, the newer ships would have been made with thinner hulls and carried fewer lifeboats. After all, the design of their new ship was based on the unsinkable, unsunk, and thus eminently successful *Titanic*.

But as we know from its colossal failure to reach New York, even the *Titanic* could not withstand an actual collision with an iceberg—a fatal flaw in the ship's design. All subsequent ocean liners whose design was closely based on the supposedly successful *Titanic* would likely also have had the same latent flaws as their paragon. In fact, because of the overconfidence in the ship's success, the inevitable use of thinner steel in their hulls would have made the derivative ships even more vulnerable, and the fewer lifeboats would have made any accident at sea potentially more tragic. Chances are, one of these "improved" ships would eventually have had the bad luck of being in the same place at the same time as a fateful iceberg. Only then might the folly of thinner hulls and fewer lifeboats, not to mention a fatally flawed bulkhead design, have become incontrovertibly evident. Successful change comes not from emulating success and trying to better it but from learning from and anticipating failure, whether actually experienced or hypothetically imagined.

Indeed, nothing about the *Titanic*'s accident, failure, and sinking should have been beyond imagination in prospect. It was well known that icebergs were likely to be encountered on the North Atlantic sea lanes, especially in the month of April, when the *Titanic* sailed. A collision with an iceberg was thus a credible accident scenario. One form such a collision could take was a grazing, in which case the ship's hull could be gashed open, or its rivets sheared off. Either way, water would be let into the bow, reducing its buoyancy. As water continued to flow in and the bow dipped farther down, the bulkheads—which only went so high, a fatal design flaw—would be overtopped. The bow would then continue to sink lower, raising the stern. If the stern were raised out of the water, a condition for which the structure was not designed, then the ship could break in two, and its sinking would shortly follow. It certainly was an easy calculation to make to determine that the number of people in distress would far outnumber the capacity of the lifeboats.

This failure scenario, which is now what is believed to have happened in fact, should have been the basis for obviating faulty design decisions. However, whether due to ignorance, overconfidence, or rationalization, neither the design of the ship nor its operation seems to have been modified or adjusted to ensure that the not-unlikely scenario did not play out in actuality. Everyone involved—from designers to owners to crew to passengers—seems to have expected success more than feared failure, perhaps owing to the generally infrequent occurrence of ships hitting icebergs on transatlantic crossings. But success is a fickle guide, and we should always want to balance our hopes for success with a proper acknowledgment that failures can and do occur. Failures, after all, provide the lessons and wisdom to foresee even beyond the hypothetical wherein a newly proposed design, plan, or policy is likely to go awry. An overreliance on past successes is a sure blueprint for future failures.

Engineering is surely a profession focused on the future, all too often looking back mainly to calibrate progress; engineers are typically drafting plans for the next generation of artifacts, seeking to design and achieve what has not been done before. Engineers are ever conceiving larger, faster, more powerful structures and systems; and they are also ever devising smaller, lighter, more efficient machines and devices. Every new thing has meaning in comparison with that which it supersedes—bettering the past, as it were, and usually a very recent past.

If previous engineering achievements have any relevance in such a climate, it would thus appear to be principally as standards against which the latest designs are judged. From this point of view, only the most recent technological history would seem to be really relevant for modern engineering, and then only insofar as it presents data to be extrapolated or a challenge to be overcome. Any temporally extended history of engineering would be mostly irrelevant technically. The history of engineering as a succession of achievements, of incremental progress, might motivate young engineers and give them and veterans alike pride in their profession, but it is not commonly expected to add directly to their technical prowess.

Focusing on the most recent past is a very shortsighted view. The long history of engineering, when embedded in a social and cultural context, does in fact have considerable potential for demonstrating the true nature of engineering as practiced in the real world today, but only if it is presented in a way that makes it engineering as well as history. For no matter how profoundly engineers know that their problems have more than a technical dimension, their solutions ultimately will suffer if the engineers do not base them on solid technical ground. Thus one of the most promising uses of history in engineering education and practice is to add to a fundamental understanding not of how obsolete artifacts worked but of the timeless aspects of the engineering process itself, while at the same time providing an appreciation for the past and process of civilization itself and engineering's role in it. But as everyone knows, the past of engineering is characterized not only by successes but also by failures.

In an open discussion of the relevance of history held before the British Institution of Structural Engineers, R. J. M. Sutherland expressed the opinion, which was echoed frequently by the discussants, that major engineering disasters "are much more likely to be avoided if future designers, individually, develop a habit of looking back and questioning how each concept grew." Unfortunately, this is seldom done. Yet ironically, signal successes in engineering have tended to arise not out of a steady and incremental accumulation of successful experience but rather in reaction to the failures of the past—from the minor annoyances accompanying existing artifacts to the shock of realization that the state of the art was seriously wanting. Thus the collapse of the Tay Bridge, designed by an overconfident Thomas Bouch, led to the abandonment of his design for another bridge on the same railroad line. Rather than going ahead with Bouch's ambitious suspension bridge across the Firth of Forth, the railroad commissioned a new engineer, John Fowler, who with the young Benjamin Baker designed the monumental cantilever structure that still is in service at that crossing. The collapse of the first Quebec Bridge, which had been designed to surpass the Forth in length and economy, led to the redesigned second, now a symbol of engineering response to failure and of Canadian resolve—not to mention its inspirational role in the

Iron Ring tradition. The colossal collapse of the Tacoma Narrows Bridge taught bridge engineers overnight an appreciation for aerodynamics that in time led to such new and successful suspension-bridge designs as those across Britain's Severn and Humber rivers. An enlightening and efficacious history of engineering should incorporate a treatment of these and other engineering failures and failure-driven successes not only for their value in adding a measure of humility to the innate hubris of engineers but also for the essential features of the self-correcting engineering method that they can so effectively reveal.[1]

The history of engineering, as that of civilization itself, is clearly one of both successes and failures, and paradoxically the failures are the more useful component of the mix. Although examples of good engineering practice and grand technical successes can certainly serve as paradigms of good judgment and works to emulate, great engineers and great people generally do not become so merely by reading biographies of great men and women. And great new engineering achievements do not come to be so merely by inference from an extrapolation of successful precedents. Indeed, the history of civil engineering is littered with the wreckage of famous bridges that were designed in a tradition of success: the Dee Bridge in 1847, the Tay Bridge in 1879, the Quebec Bridge in 1907, and the Tacoma Narrows Bridge in 1940. Although each was pushing the limits of the state of the art in terms of length or slenderness—and economy—none of these structures was forging any fundamentally new technology; each seemed to be but an incremental step (albeit, in some cases, a too-large one) in what had already been done more or less successfully.[2]

One very interesting study of the failure of bridges either under construction or recently completed or modified was published some years ago by Paul G. Sibly and Alistair C. Walker. The study was based on Sibly's doctoral dissertation, in which he found that major bridge disasters that had occurred between the mid-nineteenth and the mid-twentieth centuries had happened with surprising temporal regularity—with significant failures taking place at roughly thirty-year intervals. The pattern established by the Dee, Tay, Quebec, and Tacoma Narrows bridges continued when in 1970 two steel box-girder spans—one located in Milford Haven, Wales, and one in Melbourne, Australia—collapsed while under construction. (Since Silver Bridge was four decades old when it collapsed in 1967, it does not fall into the same category. Nor does the Minneapolis bridge that collapsed in 2007, though the placement of heavy construction equipment and materials on it may be said to have made it recently modified.)[3]

The striking regularity of the occurrence of major bridge disasters led to the expectation that the pattern would be continued by a failure that would occur around the year 2000. As the close of the twentieth century approached, the most likely bridge type to fulfill the prediction appeared to be the cable-stayed. Throughout the 1990s, cable-stayed bridges had been experiencing ongoing cable-vibration problems and were being retrofitted with vibration-damping devices in a variety of ways. Among the earliest indications of problems with stay cables were observations made in Japan, where the vibrations were aggravated by the presence of rain. Cable casings began to be fitted with raised spiral strakes to break up the rivulets that were blamed for the unwanted motion. Other examples abound. The record-setting (2,808-foot main span) Pont de Normandie, located at the mouth of the Seine, upon completion in the mid-1990s exhibited cable and deck vibrations significant enough to cause it to be retrofitted with orthogonal cable ties and also with shock-absorbing dampers, all of which diminished the appearance of the sleek structure. On a visit to Sydney, Australia, in 1998, I was driven over the new Glebe Island Bridge, whose main span is about 1,140 feet long. My host explained to me that engineers had recently decided to retrofit it with dampers to check undesirable vibrations. In spite of these behavioral problems, there has been no dramatic collapse of a prominent cable-stayed bridge, though the problems persist and increasingly cable-stayed designs incorporate vibration-ameliorating features. During a 2009 visit to South Korea, I was taken to the recently completed Incheon Bridge, and it too had dampers in place to check cable vibration.[4]

In spite of these examples and more, cable-stayed bridges have continued to be built with increasingly longer spans, and cable vibration problems have become commonplace, if not expected. Whether the continued reliance on retrofitting so many of these bridges with damping devices will prevent a colossal failure in the wind remains to be seen. What is incontrovertible is that such measures have reduced dramatically the unwanted vibration problems of the structures, checking their motion to within acceptable limits. However, such retrofitting appears to have resulted in a sense of confidence within the bridge-building community that if problems arise in a new bridge they can be brought under control. This is very much analogous to the situation that existed with suspension bridges in the late 1930s—until the appearance of the theretofore unobserved torsional motion of the deck of the Tacoma Narrows Bridge led within hours to its collapse. Although through the year 2010 there had been no similar catastrophic failure of a cable-stayed structure, a sudden twist in the behavior of one might still be cause for concern.[5]

The pedestrian bridge also experienced a great spurt of creative development in the 1990s. Footbridges are nothing new, of course, having been perhaps the first bridges, and it is their very pedestrian nature—literally—that seems to have made them appear to be but modest challenges to bridge designers. Indeed, it may be precisely because of their long history and familiarity that their engineering has seldom in recent times been the focus of much attention. Architecture, aesthetics, and the use of new materials in footbridges have typically been much more discussed than their

structural engineering. This all changed at the end of the twentieth century.

Pedestrian traffic subjects a bridge to quite a different kind of
loading than does vehicular traffic or the wind. It would appear
to be a lighter and gentler load, but that does not mean that it
might not be a more difficult load for a particular bridge to bear.
Recall that on the day in 1987 when the Golden Gate Bridge was
closed to vehicles to celebrate its fiftieth anniversary, so many people crowded onto the structure that it was subjected to the heaviest load it had experienced in its lifetime. The total weight of the
people has since been much discussed, along with the resultant
noticeable downward deflection and flattening of the center of the
bridge's main span, but the presence of so many pedestrians also
caused the bridge to sway sideways. A similar sideways motion
occurred in New Zealand in 1975, when protesters took over the
Auckland Harbour Bridge. But footbridge engineers have generally considered such examples of swaying to be anomalies or irrelevant occurrences.[6]

In the years surrounding the turning of the millennium, footbridges came to be seen as more than just pedestrian, utilitarian
structures. The town of Gateshead, England, sponsored a competition for teams of architects and engineers to design a distinctive
footbridge, which resulted in the innovative Gateshead Millennium Bridge that opens for shipping along the Tyne like an eyelid
to the morning light, and closes after a ship has passed in the night.
With their penchant for nicknames and slang, the British chris-
tened the structure the "blinking-eye bridge." In Japan, the structural engineer Leslie Robertson teamed up with the architect I. M.
Pei to produce the dramatic bridge that is part of the ceremonial
entrance to the remotely located Miho Museum. In several Western countries, the engineer-architect Santiago Calatrava has designed cable-stayed footbridges dramatically supported by a single
mast, such as the span that now carries people from downtown

Milwaukee to the new wing of its art museum, also designed by Calatrava. And in London, an engineer, an architect, and a sculptor teamed up to produce the Millennium Bridge linking across the Thames the Tate Modern art gallery and St. Paul's Cathedral.[7]

The London Millennium Bridge is, of course, famous for having been shut down just days after its opening in June 2000, when the bridge deck started swaying sideways to an alarming degree. A similar thing had happened the prior year to a footbridge in Paris. The Passerelle Solferino was designed to give pedestrians a serene crossing of the Seine from the Tuileries Quay to the Musée d'Orsay. Although tested before opening with 150 people dancing to a beat designed to reveal dynamic susceptibilities, this technologically and aesthetically innovative arch bridge swayed on opening day. The one-inch movement, amplified by French politics, forced the closure of the bridge that same day. Closing bridges susceptible to swaying is a wise decision, for even just the hint of a bridge's possibly impending fall can trigger panic and tragedy. A classic example occurred on Memorial Day 1883, when a week after the Brooklyn Bridge opened a dozen people were trampled to death on its walkway when holiday strollers panicked about the safety of the structure. In 1958, a pedestrian suspension bridge in Kiev, Ukraine, was closed because it swayed a fraction of an inch when crowded on the weekends. More recently, in November 2010, when millions were in Phnom Penh, Cambodia, for an annual water festival, a swaying suspension bridge of modest span was the scene of the death of 350 people who were crushed or suffocated; another 400 people were injured in a stampede that was believed to have been set off by fears that the structure was unstable. Unfortunately, the bridge was not closed until after the disaster occurred. The architect-engineer of the Passerelle Solferino, Marc Mimram, admitted that designing "a footbridge is more difficult than other bridges, with its conflicting demands of light weight and long span." Still, how could such embarrassing oversights happen, especially at the dawn of the twenty-first century, when engineers were using powerful, computer-based tools to design what their ancestors once sketched out with nothing more than sticks in the sand?[8]

As with the other bridge types that Sibly had studied, the design of footbridges had become routine. Unfortunately, the loads considered in their design did not include the sideways forces exerted by people walking, forces that have a frequency of one-half that associated with a person's vertical footfall. Historically, with bulkier bridges, this had not been a problem, but the natural frequency of sideways motion of the slender Passerelle Solferino and the London Millennium Bridge was close to the frequency of sideways force that pedestrians exert in the course of moving in their normal gait. Though crowds of people do not generally walk in step, if the bridge beneath them begins to move sideways—for whatever reason—the people on it instinctively tend to fall into step the better to keep their balance. This in turn exacerbates the sideways motion of the structure, and a positive feedback loop is developed. The motions can get so violent that public safety concerns dictate the closure of the bridge.

The Passerelle Solferino and London Millennium Bridge do not look anything alike. The former is an arch and the latter a low-profile suspension bridge. Yet although they do not look similar, they shared with other footbridges fundamental design assumptions, which clearly did not take into account the critical sideways loading mode. In this sense the development of footbridges falls into the pattern pointed out by Sibly, namely, that the state of the art, which had developed out of successful experience, was finally being pushed, albeit inadvertently, into realms for which it was inadequate. A previously ignorable characteristic of the design came to dominate structural behavior. By this criterion, the failure—even though not a dramatic collapse—of these footbridges, in which dynamic phenomena insignificant in bridges of lesser magnitude revealed themselves to be limiting, would appear to have continued the thirty-year cycle and fulfilled the expectation of a bridge failure around the year 2000.

But this has not stopped the design of ever-more-daring crossings. Just as vehicular suspension spans were being made longer and more slender in the decade leading up to the collapse of the Tacoma Narrows Bridge, so pedestrian bridges have been getting longer and more slender. This trend will undoubtedly continue in the absence of an actual catastrophic failure resulting from sideways sway. The 1,000-foot-long David Kreitzer Lake Hodges Bicycle and Pedestrian Bridge in San Diego, California, has 330-foot spans that are only sixteen inches deep. Thus this so-called stress ribbon bridge, essentially a suspension bridge with the deck resting directly on the taut cables, has a span-to-depth ratio of 248, compared with 350 for the doomed Tacoma Narrows. Another San Diego footbridge, connecting a park with the convention center and "designed as an iconic gateway into the city," ran a year behind schedule and in the end cost more than double the original estimate. The bridge was described as "a one-of-a-kind structure," which can explain the delay and cost overrun. When concrete was poured for the bridge's deck, it was found to be 7 percent heavier than planned, which required changes to the support cables. Such complications may have occurred in a more conventional design, but iconic structures tend to invite unique problems and complications. Under the best of circumstances, those complications can actually make engineers step back and rethink their designs, which in turn can lead them to uncover potential flaws and so correct them before failure occurs. However, not all warning signs of failure are heeded and, in fact, the most ordinary aspect of a project can lead to trouble.[9]

But why, if they had been in use and under development for so long, did footbridges—and in particular the London Millennium Bridge—fall so neatly into the thirty-year failure cycle? Apparently, major bridge failures have occurred roughly once a professional generation, which is typically about three decades long. Depending upon whether a new or significantly extended design is undertaken close on the heels of a major failure or some years after it, engineers of all kinds and ages may or may not be sensitized to their fallibility. Also, in the course of a long period of success, they may become complacent in the practice of their art. According to Sibly and Walker, writing in 1977,

> The accidents happened not because the engineer neglected to provide sufficient strength as prescribed by the accepted design approach, but because of the unwitting introduction of a new type of behaviour. As time passed during the period of development, the bases of the design methods were forgotten and so were their limits of validity. Following a period of successful construction a designer, perhaps a little complacent, simply extended the design method once too often.

In the case of the London Millennium Bridge, the new behavior was the sideways sway of the structure induced by the sideways horizontal forces exerted by pedestrians on an unusually shallow suspension bridge. The sway was amplified when the pace of the walkers happened to coincide with the natural frequency of the structure. Bridge designers had long concerned themselves with the vertical footfall of pedestrians, with some nineteenth-century bridges having had posted warnings to soldiers to break step when using the crossing. However, the horizontal component of their footfall had had little or no effect on more conventional structures, and so it had become first something ignored and then something lost to the design consciousness.[10]

The problem is exacerbated when the engineer is a generation removed from the inception of a design method or structure type.

Without a knowledge of the fundamental design assumptions that were made during the development stage—assumptions that implicitly if not explicitly limit the applicability of the method or structural type—younger engineers can be effectively designing blind in uncharted territory, all the while thinking they are just following a well-worn path. The phenomenon is not unique to bridge engineering.

The use of steel in building construction dates from the 1880s, when the first skyscrapers were being erected in Chicago. Framing a tall building in steel columns and beams, between which concrete floors and masonry walls were inserted, made for a relatively light and efficient structure. The steel was also encased in concrete or tiles, thus providing a degree of fireproofing. (Unlike the Twin Towers of New York's World Trade Center, which collapsed in the fires that burned on September 11, 2001, an almost century-old building nearby at 90 West Street survived precisely because of its concrete-, brick-, and terra cotta tile-encased steel columns.) In the early use of steel in buildings, beams and columns were connected by means of heavy riveted joints, but over time these evolved into less and less stiff designs. Furthermore, whereas older steel-framed skyscrapers, including the Empire State Building, gained stiffness from the infilling and encasing masonry and concrete walls, post–World War II structures did not, because their glass façades, known as curtain walls, were merely hung from the structure's frame. In the meantime, welding had replaced riveting in making connections. All this evolutionary development did make for more efficient and economical structures, but ones that had latent flaws.[11]

These flaws were revealed during the Northridge earthquake, which struck southern California in 1994. Structural engineers were surprised by the number of brittle fractures that appeared adjacent to welded connections between beams and columns. A similar phenomenon occurred the following year, when a large earthquake struck Kobe, Japan. Needless to say, the unexpected structural behavior prompted investigations, and these found that the fractures were attributable to a combination of factors, all of which may be said to have developed innocuous-step-by-innocuous-step over the course of the preceding century. Individually, none of the changes in the long transition from robust riveted to spare welded connections might have led to the failures, but collectively they all contributed to the surprising result. Among the contributing factors were the loose standards imposed upon the steel used, the use of weld metal that had a low tolerance for imperfections, poor welding practices, and insufficient quality control on work done. It was a classic case of devolution from a robust design that was incrementally eroded in the name of economy and efficiency. Needless to say, new standards were instituted in the wake of the Northridge and Kobe earthquakes, something that might not have been necessary if the fundamental considerations that went into the oldest designs were retained in creating the newest ones—in other words, had there been more of an institutional memory and awareness of historical development within the profession of structural engineering.[12]

The National Aeronautics and Space Administration, whose engineers and managers think about and design space missions that can literally take generations to come to fruition, has focused on "filling the knowledge gaps." This has involved a conscious effort to write reports that document more than just results. The new reports are "vastly more detailed" than older ones, because veteran engineers "do not want future researchers to have to try and figure out what steps we took and why we took them." There was a sense at NASA that as it undertook "ambitious new programs with a new generation of engineers and managers, it was more important than ever to make sure that valuable experience-based knowledge gets passed from project to project and from an older generation to a new one."[13]

Even within a technical generation, problems can arise due to infrequent exercise of skills and procedures. The Taurus XL rocket was a "beefed up model" of a rocket that was capable of launching small satellites into orbit. In 2001, an earlier version of the Taurus failed to achieve orbit when a steering mechanism jammed; the XL was equipped with a redesigned steering system as well as larger rocket stages that boosted its payload capacity to 3,000 pounds. Its first launch, in May 2004, was a success, but the launch vehicle was used sparingly over the next five years. In fact, overall only eight Taurus rockets were employed, with launches typically taking place several years apart. In February 2009, the Orbiting Carbon Observatory, whose function was to make precise measurements of atmospheric carbon dioxide, was lost when the Taurus XL carrying it into orbit failed to do so because the fairing protecting the satellite during rocket flight did not separate properly, thereby preventing the vehicle from reaching its target speed and altitude. The rocket and its payload fell back to Earth and settled to the ocean bottom near Antarctica. Even without the benefit of malfunctioning parts to analyze, some hardware problems were identified as possible "intermediate causes" of the failure, but the root cause was believed to be "the organizational behaviors, conditions, or practices that ultimately led to the production and acceptance of what proved to be faulty mechanisms." Fixing the intermediate causes without addressing the root causes could only be expected to lead to future failures. The leader of the NASA investigation board that analyzed the Taurus XL case concluded his recounting of the investigation with a perceptive observation: "Many of the people involved with launching the Orbiting Carbon Observatory had little or no experience with the launch vehicle. The less often you launch, the more attention you should pay to the formal procedures that embody much of the information and knowledge past practitioners have acquired about how to launch successfully."[14]

And launching successfully means, of course, anticipating potential modes of failure and designing against their occurring. However, without experience or knowledge of past failures, or, equivalently, with nothing but positive experiences of successes, we are at a great disadvantage. Formal procedures may embody lessons learned, but engineers and managers alike are notorious for deviating from past procedures. After all, we are always seeking to improve on the past, and this means changing the way we design, make, and do things. Unfortunately, in the course of making changes, we can effectively discard the experience embedded in that which we think we are improving upon. For this reason we can, ironically, expect failures to continue to take us by surprise.

We should not assume that the thirty-year pattern of bridge failures has ended with the dawn of a new millennium. The problems with the London and Paris footbridges provided a renewed sensitivity to the possibility of failure even in structures designed according to the latest engineering knowledge with the latest engineering tools. In particular, that knowledge and those tools have limitations, and until those limitations are discovered—most likely through some unexpected behavior or an outright failure—we can expect to be surprised. So which bridge types are most likely to be designed or developed in an atmosphere of incomplete knowledge, and can we expect a dramatic failure of one of them to provide the circa 2030 data point that will continue Sibly's thirty-year cycle? My educated guess is that it will be either a cable-stayed bridge, since engineers continue· to extend that form into uncharted territory even as the unwanted vibration of its cables continues to baffle them, or a post-tensioned concrete box-girder bridge, an increasingly familiar form whose cantilever construction methods are pushing the limits of its applicability.[15]

Although Sibly's original thesis was prompted by a study of metal bridges and strictly speaking applied only to them, its applicability to a wider variety of designed structures was soon apparent. Thus he and Walker generalized from the thirty-year cycle of bridge failures:

> Our studies show that in each case, when the first example of a technologically advanced structure was built, great care and research went into its design and construction. But as the new design concept was used again and again, confidence grew to complacency and contempt for possible technical difficulties. Testing was considered unnecessarily expensive and so it was dispensed with. But in each case, the design was steadily modified and the changes not understood, until a previously ignored second order effect dominated, and the structure collapsed.

In this case, Walker and Sibly were writing in 1976 in the wake of a meeting on offshore oil platforms, at which "top designers in the field admitted that they do not understand the dynamic forces the North Sea puts on their platforms," whose technology was "being extended far beyond the validity of simple design tests." They might just as well have been writing in 2010 about the design and testing of oil rigs and blowout preventers used in the Gulf of Mexico.[16]

Such observations fly in the face of the conventional wisdom that technological knowledge is constantly advancing. But if technology itself is cumulative, with each incremental improvement adding to our store of engineering knowledge and achievement, then why shouldn't all newer bridges and oil rigs be better than their predecessors? In particular, why should any ever fail? And why shouldn't a younger generation of engineers know more and be wiser than the older? And how then can cyclic behavior—with decades of success being punctuated with colossal failures—be explained? The explanations lie in the nature of engineering design, which begins in a most primitive and nonrational way, and for which previously untried designs almost always start from scratch. Engineers literally dream up designs, and more often than not in nonverbal, graphical form. Only when an engineer has the equivalent of a sketch or drawing can unambiguous conversation take place among engineers and designers and can the engineering sciences, economics, and expert experience be called upon to analyze and judge the practicality and constructability of a concept. The design process is often convoluted and iterative, of course, and an engineer's sketch can be informed, consciously or unconsciously, by a wealth of knowledge of engineering and science and of experience with past designs, both successful and unsuccessful. Such information and insight help the mind's eye select and reject features of a tentative design. If this is done in ignorance of past failures, there might be nothing to trigger an alarm that history may be about to repeat itself.

In *Engineering and the Mind's Eye*, the historian of technology Eugene Ferguson elaborated on the role of nonverbal thought in engineering and put it in a historical context. Although his work demonstrated that there was a definite evolutionary pattern in the quality and clarity of the graphic representation of artifacts and ideas for artifacts, the unmistakable lesson was that the engineering design process does not proceed in modern times in any fundamentally different way from how it did millennia ago. Egyptian pyramids, Roman arch bridges, Greek temples, and every other engineering project of ancient civilizations began in the mind's eye of some individual engineer, whether or not the individual was so designated, and whether or not he could draw as well as his descendants. Regardless of drawing capabilities or talent, it was only when a conceptual designer articulated his idea by a sketch in the sand, a drawing on paper (or papyrus or wood), or a model in clay that it could be praised or criticized by his colleagues, accepted or rejected by the commissioners, or built or bungled by the contractors.[17]

For all of engineering's real progress in technical knowledge and analytical technique, then, the underlying nature of design is fundamentally the same today as it has always been, and therein lies the true potential of generational knowledge transfer or of a properly conceived engineering history, or even of just an anthology of case studies: such an approach can not only bridge the gap between the nontechnical and the technical aspects of engineering practice but also provide a theoretical framework within which the nature of engineering itself can be articulated. Furthermore, the wealth of wisdom about the seemingly ineffable engineering method that exists in the minds of older engineers and in countless classical case studies can then be brought out of the library, into the classroom, and, ultimately, into the design office. Good war stories about the engineering design process are never obsolete, for they reveal the timeless nature of engineering, warts and all.

The written record of nineteenth- and early twentieth-century engineering contains quite explicit and detailed discussions of great projects (such as the innovative Britannia, Forth, Niagara Gorge, Eads, Brooklyn, Hell Gate, and George Washington bridges, to name but a few) that tested the analytical and construction capabilities of their time. Perhaps precisely because of the tradition of writing about and explaining them in detail, the engineers of such projects tended to be more explicit about what motivated, drove, and checked their designs than are engineers today, whose reasoning can be lost amid mathematical and computer models, codes, graphics, and printouts. During a discussion of the relevance of history to structural engineering, one participant confessed that he found it somewhat easier to get into the heads of designers from earlier times than to do so with design teams typical of the computer age. Some features of engineering design will forever remain independent of the state of the art and, therefore, are independent of whether we use abacuses, slide rules, or digital computers to analyze and flesh out the conceptual designs from which all engineering flows. Even as the design process comes under increasing scrutiny, especially in a digital context, more temporally distant writing about engineering remains one of the most illuminating sources of insight into the often ineluctable engineering method.[18]

The legendary foundation engineer Ralph Peck wrote widely and articulately on engineering judgment, that attribute generally thought most difficult to teach and learn. In discussing the reliability of dams, Peck observed as late as 1981 that "nine out of ten recent failures occurred not because of inadequacies in the state of the art, but because of oversights that could and should have been avoided." Peck also pointed out that, when it comes to design errors and failures, "problems are essentially nonquantitative" and, furthermore, that "the solutions are essentially nonquantitative." He acknowledged that improvements in analysis and testing might certainly be profitable, but it is also likely that "the concentration of effort along these lines may dilute the effort that could be expended in investigating the factors entering into the causes of failure." Among Peck's prescriptions for restoring good judgment to engineering practice was a historical perspective; he deplored, for example, the fact that individuals taking the examination leading to registration as a structural engineer in Illinois could not properly identify such significant professional achievements as the completion in 1874 of the Eads Bridge across the Mississippi River at St. Louis.[19]

The practice of engineering is literally as old as civilization, and indeed civilization as we know it is hard to conceive of without the work of engineers, regardless of what they may have been called before modern times. But whereas many of the classic works of civilization, from poetry to pyramids, have long been taken on their own timeless terms, the methods of engineers are often thought to have been continually superseded by new, improved schemes. Although helicopters may have overtaken ramps for placing pyramidal capstones, and computers may have made possible designs beyond the calculational reach of engineers of only a few generations ago, there are fundamental aspects of the conceptual engineering design process that have changed little (if any) over millennia. In fact, the ready availability and power of the tools of modern technology may even threaten to erode the more basic conceptual engineering skills. Ironic as it may seem, we might find an antidote to the erosion of fundamental design skills and critical engineering judgment more readily in some of the oldest volumes on engineering than in the most modern textbooks.

When my university library still maintained a branch in the engineering school, I often would walk over to it after lunch and read the latest journals or browse through its book stacks. Current periodicals were located on the main floor, bound volumes of back issues were located one floor down, and books and monographs were one floor up. It was possible to read the latest articles without having to use the stairs, but I often did so in order to roam about the old volumes and see where serendipity would take me. In the bound-periodicals section, titles were arranged in strict alphabetical order, and the latest volume of a magazine or journal was naturally placed chronologically at the rightmost end of the run of that title. A space was left between that last volume and the beginning one of its alphabetical successor, so that more recent volumes could be fit in without having to adjust the positions of a lot of other volumes on the whole range of shelves.

On the books-and-monographs level, however, all titles were arranged strictly by subject matter, shelved according to the Dewey Decimal System, which my university's library then still used. Because the shelves were nearly filled to capacity, it was uncommon to find much empty space among the books shelved cheek by jowl. Still, this arrangement seemed ideal to me, for it fortuitously juxtaposed sometimes the newest with the oldest scholarship—and all in between—on a specific topic. If I went to a section on bridges, I might find a nineteenth-century monograph on suspension-bridge design shelved next to a late twentieth-century one. Drawn by serendipity to such a pairing, I might take both to a nearby carrel and find in the old and in the new fundamental similarities in conceptual approach even though their analytical and graphical content showed them clearly to be of different eras and types.

One thing that especially impressed me was that, seemingly no matter when a book on design was published, it discussed either explicitly or implicitly the persistent problem of failure and how to avoid it. The books often contained case studies of significant failures, sometimes in an introductory chapter, sometimes in an appendix, sometimes in the body of the text, and sometimes spread throughout the book. Famous failures might just be mentioned in passing, or alluded to by geographical location only, the writer obviously expecting the reader to be quite familiar with and well-versed in their stories and lessons. Seldom did I come across a book that did not acknowledge, albeit sometimes only implicitly, that failures had occurred in the relevant field and that knowing the circumstances and understanding the causes were valuable pieces of information and knowledge that provided perspective and perception—and experience by proxy.

One day, after reading a book review over lunch in my office, I walked over to the engineering library to look up the recently published title. I had gotten its call number from the online catalog and so went directly upstairs and over to where I knew the volumes with that number on their spine were shelved. I was disappointed not to find the book I was after, and I thought that perhaps someone had beaten me to it and already checked it out, although there was no gap in the row of books on the bookshelf to indicate that was the case. I went downstairs to the circulation desk and asked the librarian about the status of the book. He took the slip of paper on which I had written the call number and looked it up in the circulation files. According to them, the book was not checked out, and so he went upstairs with me to see if perhaps I had looked in the wrong place or if the volume had been misshelved somewhere close by to where it should have been. He could not find the book, and as we walked back downstairs he asked me its title.

Since I was such a frequent patron of the engineering library, its librarian was very well aware of my reading habits. He knew that they tended toward older rather than newer books, and so he had naturally assumed that what I was looking for was in the former category. Upon finding the book in the catalog, however, he saw that in this case I was looking for a book published just the previous year. It was only then that he made me aware of a new book-shelving policy that he had instituted at the request of some of my colleagues in the relatively new field of biomedical engineering. Since they were seldom interested in books published more than five years ago, they had little desire to negotiate their way among older titles—even though they were comparatively few in this relatively new field—to locate a recently published book. Furthermore, the older books could cause two newer ones on closely related topics to be too far apart for the engineers' liking.

The new policy addressed this complaint by adopting the biomedical engineers' suggestion: shelve books newer than five years old separately from the rest of the collection. Newer books were now grouped together at the beginning of the range of book stacks, right beside the stairwell. This would necessitate culling the newly-turned-five-year-old books every year, but that inconvenience was considered acceptable to the librarian. Perhaps he had not told me he was even considering this bibliographical age-discrimination because he knew that I would object. As soon as I learned of it, I did express my displeasure. But now that I knew where the book I was looking for was located, I went back upstairs to find it. When I reached the newer-books shelves, I was struck by the extent to which computer-science titles dominated the collection. Fully half of the newer books, at least, seemed to be on this subject. By contrast, I found that the book I was seeking had relatively few neighbors on its subject, and I could take them all in at a single glance and carry them in a single armful to a nearby carrel. So there were advantages to having all the newer books on the topic close at hand, but I worried that for all that was gained in convenience there would be a net loss in wisdom.

With the increasing sophistication of online library catalogs, book and article delivery services—not to mention Google book scans and searches—and other digital and electronic conveniences, fewer and fewer faculty members continued to use our engineering branch library, and students had largely abandoned its well-worn tables and chairs in favor of the newer and more comfortable study furniture in the recently opened addition to the main library. Even I began to find the digital resources on the Internet more user-friendly and inviting—not to mention a different but also fertile source of serendipity—and so I had less and less need to visit physically any bricks-and-mortar library on campus. Without clientele, the engineering library (and other specialist branch libraries) had an increasingly difficult time defending and occupying the space that the engineering school was kept from expanding into, and so its collection of books and periodicals was moved into the main library or into storage in a remote location. There was no protest to speak of.

It will be interesting to know if the separation of books, first from their older generations, and then from our physical contact—and the temporal and physical fragmentation of knowledge that it symbolizes—will lead to an increase in failures. The computer science and software engineering fields are still relatively new by civil engineering standards, and some of their practitioners early on began to realize that they did not have a very deep history from which to draw general lessons, especially regarding failures and cycles of failures. Software designers were particularly aware of the fact that the products of their work were prone to bugs and worse, and computer scientists began to look outside their field for some historical perspective on failure. They found it in bridge engineering, whose modern history of employing metal structures dates back to the late eighteenth century and whose record of landmark and even lesser failures has been well-documented and analyzed. Thus it became not uncommon to find in software engineering publications interviews with bridge and structural engineers, in which history and failure played a prominent role. Also, engineers who had studied and written about failure from a historical perspective began increasingly to receive invitations to speak at software engineering workshops and conferences. Members of the emerging field were very cognizant of what they did not know and could not know easily from within their own field alone, and so they sought historical perspective and guidance by analogy. The software engineering community adopted the practice of reporting and analyzing failures in its own and closely related fields, and contributions to the compilation in a newsletter seemed to grow with each issue and eventually went online, where they did not have to compete for space. The failures that software engineers have avoided by following the chronicle of missteps is generally not recorded—nor perhaps even recognized—but it is difficult to imagine that there have been none.[20]

In their enthusiasm for advancing the state of the art by pushing the limits of cutting-edge technology, some engineers do not look back at the history of their field. They do not see it as relevant to the work at hand. Even if they have an armchair interest in the history of whatever it is they are currently engaged in, they tend to compartmentalize that interest or see it as an avocation that they will pursue more avidly in their retirement. They see it at best as a cultural adjunct to what they are doing technically in the prime of their professional life.

This very well may have been the situation among suspension-bridge engineers, especially in the 1920s and 1930s. During that period, in a remarkable technological leap, Othmar Ammann designed the George Washington Bridge to have a main span almost double that of the existing record holder. The daring of this design—given that a potentially disastrous scale effect had been known at least since the time of Vitruvius—plus the uncommon shallowness of the extremely long deck of Ammann's bridge, provided a paradigm for subsequent suspension-bridge designers to follow at their peril. Because Ammann and his design team were working so far outside the envelope of experience, they took special care to make very explicit their assumptions and the implications of their calculations—especially to themselves. Because of this heightened awareness of being engaged in an unprecedented design effort, the engineers had to pay special attention to details and anticipate how they might lead to a failure. No doubt at least in part because so much careful attention was paid to its design, the George Washington Bridge was a resounding success. Throughout the 1930s, subsequent suspension bridges that were designed and built, with their designers selectively adopting the George Washington's successful features and apparently feeling no strong need to worry about failure, proved to be embarrassments. The result was bridges with lighter and lighter and less and less stiff decks that surprisingly moved a great deal in the wind, culminating in the Tacoma Narrows Bridge's tortured collapse.[21]

While the undisputed leaders in the field of bridge design were following Ammann's example, they were also disregarding the history of suspension bridges and their failures. Although they knew about the susceptibility to the wind of wooden-decked bridges of a century earlier, they considered that irrelevant to the design of modern steel structures. Even as they took historical exemplars as aesthetic models of how a bridge should look, they ignored the engineering lessons learned in their demise. Thus twentieth-century engineers remembered and admired Thomas Telford's 1826 Menai Strait Suspension Bridge as an aesthetic near-ideal, but they seemed to put out of mind the fact that the bridge's lightweight wooden roadway had been repeatedly damaged in the wind, as had that of many of its contemporaries. That was considered an irrelevant fact of history in an age of steel.

Yet by 1841 no less an engineer than John Roebling had distilled from his study of early nineteenth-century failures what he needed to do to design a suspension bridge that could stand up to the wind, which he considered to be the form's "greatest enemy." He concluded and explicitly stated that to be successful (that is, to not fail) the deck of a suspension bridge must have "weight, girders, trusses and stays." In other words, the successful deck must: (1) be heavy, which gave it substantial inertia, so that its mass was not easily moved by gusts of wind; (2) have girders and trusses, so that its deck possessed sufficient stiffness to maintain its shape while distributing the load it carried; and (3) have supplementary cables, or "stays," to check unwanted motion before it could get out of hand. Roebling incorporated these features first into his 1854 Niagara Gorge bridge—the first suspension bridge to carry railroad trains—then into his Ohio River bridge at Cincinnati, and finally into his masterpiece, the Brooklyn Bridge, which was completed in 1883. Unfortunately, in subsequent decades other engineers systematically eliminated from early twentieth-century suspension bridges what Roebling considered the essential features for success. The process culminated in the bridges of Ammann and his contemporaries in the 1930s. First, the cable stays had been done away with as redundant, leaving those on the Brooklyn Bridge to be seen as a somewhat quaint but distinguishing architectural feature only. Next, the stiffening truss had been abandoned in the name of the aesthetic goal of slenderness. Finally, weight itself had been shed in supposedly optimally designed bridges with very narrow roadways. The culmination of this historically ignorant structural evolution was the Tacoma Narrows Bridge.[22]

In the immediate wake of the Tacoma Narrows collapse, all the lessons of history seemed suddenly to be recalled, but by then it was too late. Before long the official failure analysis of the Tacoma tragedy itself reached conclusions that did not differ all that much from what Roebling had concluded a century earlier, namely, that wind was the enemy of suspension bridges. After an understandable hiatus in the 1940s, caused by war as much as by the infamous failure, suspension bridges began to be built once again, but their wind-resisting features were more influenced by historical than by modern examples.[23]

If engineers have ignored history, historians have been guilty of ignoring engineering. In their analysis of the past, they generally have not paid as much attention to developments in technology as they have to those in politics and other areas. Yet an engineering achievement like the Brooklyn Bridge, in making commuting between Brooklyn and Lower Manhattan more convenient and predictably reliable, played a significant role in the consolidation of once-separate political and geographical entities into the New York City that exists today. For some time, historians of technology have called for a greater recognition of the effects of such technological developments on the course of human events. In the absence of textbooks giving what they considered a balanced view of history—meaning a consideration of the technical equally with the economic, the social, the cultural, and the political—some sympathetic historians teamed up to write a book that did. Perhaps this should inspire engineers to incorporate relevant history into their own textbooks. But as they have done for millennia, engineers design the future and historians analyze the past. These oversimplifications may highlight some fundamental differences between the practice of engineering and that of history, but when taken as indicative of divergent and exclusionary objectives they can lead to both inferior engineering and incomplete history.[24]

Stories of success and failure and their lessons can illuminate more than just engineering and history, of course. They are also relevant to current affairs. The financial crisis that began in 2008 provides a cogent example. Prolonged appreciation (success) in the housing market promoted the irrational feeling that real estate could and would appreciate indefinitely. At the same time, past (successful) mortgage practices were stretched to encourage home buying beyond buyers' means. Long-standing rules of thumb, which incorporated what engineers would term factors of safety against failure and regarding what portion of one's income was wisely committed to paying off a mortgage, were ignored. The market for risky financial instruments depended on never-ending favorable real estate and economic conditions. But, of course, everything has its ups and downs, its successes and its failures. The financial crisis might have been more widely foreseen had there not been such a myopic focus on an ever-expanding bubble of success into the future and an aversion to thinking about the possibilities of failure. Recall that it was two years after the height of the financial crisis that the chairman of the Federal Reserve Bank finally admitted that he had failed to recognize the deeply embedded "flaws and weaknesses" in the financial system. Past successes may be inspirational and encouraging, but they are not by themselves reliable indicators of or guides to future success. The most efficacious changes in any system are informed not by successes but by failures. The surest way for the designer of any system to achieve success is to recognize and correct the flaws of predecessor systems, whether they be in building codes or in banking policies or in bridges.[25]

The history of civilization itself has been one of rises and falls, of successes and failures. Some of these have been of empires, dynasties, families; others have been of nations, states, and cities.

Common to all of them has been the human element, embodied in the ruler and the ruled alike. Given that the ultimate unit of civilization is the individual, we need look no further than within ourselves to gain insight into the world and its ways, including the failure of its institutions and its systems. And, just as those institutions and systems are made up of individual people and things, so ultimately must we look to ourselves and to how we interact with the world, both given and made, whenever something goes wrong.

None of us should be able to say that we have succeeded at everything we have tried, nor can we say that we have done nothing but fail. Even the most ebulliently optimistic among us must admit, if only in their private thoughts, that they do not have a secret formula for success; and the most lugubriously pessimistic must know that they have learned from their missteps where the surest footing lay. We all ride the trajectory of our life as if on a roller coaster, with its ups and downs and swerves and curves and the occasional loop-the-loop. Some of us choose to stay pretty close to the straight and narrow track, seldom deviating from the gently rocking rides of the kiddie parks of our childhood; others of us are ever the thrill seekers, making pilgrimages as it were to the latest amusement park sensation that pushes the limits of structure and courage alike, always reaching for the sky and taking the downhill slides in stride. Most of us are somewhere in between, of course, working our way slowly and deliberately upward over higher and higher crests—until losing our lunch on the way down from the top.

Understanding failure is as easy as understanding life, even if we live a life of clichés. We know, if only from watching sports on television, both the joy of victory and the agony of defeat. We know, if we just allow ourselves to know, that when we are on a roll we almost invariably get moving too fast for our own good. Pride does indeed come before the fall. We also know, mostly from the school of hard knocks, that when we are down we must just get ourselves up, dust ourselves off, and start all over again.

Even though conventional wisdom has it that engineers are a different breed of people, they are inside just as human as everyone else. Engineers are in fact as likely to succeed and fail in life—and on roller coasters—as the next guy or gal. But if they are indeed just ordinary people, how can we expect their creations—their designs—to have qualities their creators do not? Fortunately, engineers often work in teams, whether actual or virtual, checking and catching one another's ideas and calculations for false dreams and true errors. The ups and downs of one engineer are typically balanced by the downs and ups of another, and this produces the safety and reliability that we know to be a hallmark of technology. A failure does happen now and then, of course, often because everyone is looking in the same wrong direction—up. This is particularly likely to happen when things have been going swimmingly and no one thinks that there is anything to worry about. The finely tuned machinery is taking care of itself. But when the crash of the roller coaster cars occurs, it provides the wake-up call to collect the pieces, scrutinize them for design flaws and fatigue cracks, inspect the rails for deficiencies, question the maintenance procedures and the operations manuals, understand the problem, and go back and redesign the ride.

None of the failures described in this book was necessarily predestined to happen. Had one of the eyebars in the St. Marys Bridge, the upriver structural twin of Silver Bridge, been heat treated in a more deleterious way or installed with a more serious undetected flaw, it may have been the one that failed first. And it would then likely have been Silver Bridge that was disassembled as a precautionary measure. Or had anyone known that a dangerous crack might have been growing undetected in one of the Silver Bridge eyebars, the extreme step might have been taken of closing the span to traffic until a definitive inspection could be conducted.

Given that the design of the suspension chain did not allow for that even with state-of-the art equipment, an alternative measure might have been to limit traffic on the structure, much as was done with the Waldo-Hancock Bridge in Maine when its suspension cables were found to be deteriorating at an accelerating rate. This would at least have bought time, during which alternative measures might have been taken. No individual failure absolutely has to happen.

The I-35W bridge in Minneapolis might not have collapsed when it did had the construction materials and equipment not been placed upon it when they were. But perhaps another deck-truss bridge of similar design, on another busy interstate highway, with another maintenance schedule might have experienced a more benign failure, which in turn might have initiated more intense and careful inspections of all steel-truss bridges, including the one in Minneapolis. And then the bent gusset plates of the I-35W span might have been looked at with a scrutiny they had theretofore escaped. Even the Tacoma Narrows Bridge collapse might not have occurred had consulting engineer Moisseiff not lengthened and made more slender the design of bridge-authority engineer Eldridge, or had the alarm sounded by consulting engineer Condron to the Reconstruction Finance Corporation not been dismissed, effectively by fiat. But if the Tacoma Narrows Bridge did not fall, then some other one designed according to the aesthetic fashion of the late 1930s would likely have—or at least would have behaved so badly in the wind as to force a reassessment of the situation.

Specific failures occur when they do because of the coincidence of any number of factors that happen to converge at a particular place and a particular time, much the way the *Titanic* had its chance encounter with the iceberg in the North Atlantic. But it is the nature of the human and technological condition that until incontrovertible failures do occur there is the tendency—even among designers, who should know better—to think that the technology has been mastered. It is not a logical conclusion, nor is it one supported by the history of engineering. It is simply the way things are.

But things need not be this way. We do have ample evidence of this irrational sense of overconfidence engendered by prolonged success, of which we are disabused only by a colossal failure. Even the gentle warnings of near-failures are seldom sufficient to shake us out of our complacency, as demonstrated so dramatically and conclusively by the space-shuttle accidents that occurred within a managerial culture of denial. But perhaps the recognition of these patterns, reinforced by failure after failure, and drummed into students and practicing engineers (and, one can hope, managers) alike by failure analysis after failure analysis, by case study after case study, will eventually bring us to shift from a success-reinforced paradigm to a failure-averse one. Perhaps then we will finally appreciate that the best way of achieving lasting success is by more fully understanding failure.

NOTES

1. By Way of Concrete Examples

1. Nolan Law Group, "Did Regulatory Inaction Cause or Contribute to Flight 3407 Crash in Buffalo?" Feb. 16, 2009, http://www.nolan-law.com/did-regulatory-indiffeerence-play-a-role-in-icing-crash.

2. Ibid.; Matthew L. Wald, "Recreating a Plane Crash," *New York Times*, Feb. 19, 2009, http://www.nytimes.com/2009/02/19/nyregion/19crash.html.

3. Jerry Zremski and Tom Precious, "Piloting Caused Flight 3407's Fatal Stall," *Buffalo* (New York) *News*, Feb. 3, 2010, http://www.buffalonews.com/home. story/943789.html; Wald, "Recreating a Plane Crash." For a perspective that sees a crew member's failure to challenge a superior as a problem rooted in culture, see Malcolm Gladwell, *Outliers: The Story of Success* (New York: Little, Brown, 2008), especially chap. 7, "The Ethnic Theory of Plane Crashes."

4. Nolan Law, "Did Regulatory Inaction Cause"; Wald, "Recreating a Plane Crash"; Matthew L. Wald and Christine Negroni, "Errors Cited in '09 Crash May Persist, F.A.A. Says," *New York Times*, Feb. 1, 2010, p. A14.

5. George J. Pierson, "Wrong Impressions," letter to the editor, *Engineering News-Record*, Feb. 11, 2008, p. 7.

6. William J. Angelo, "Six People Indicted for Roles in Alleged CA/T Concrete Scam," *Engineering News-Record*, May 15, 2006, p. 18; William J. Angelo, "I-93 Panel Leaks Plague Boston's Central Artery Job," *Engineering News-Record*, Nov. 22, 2004, p. 13; Katie Zezima, "U.S. Declares Boston's Big Dig Safe for Motorists," *New York Times*, April 5, 2005; "'Big Dig' Leak Repairs to Take Years," *Civil Engineering*, Oct. 2005, p. 26; Sean P. Murphy and Raphael Lewis, "Big Dig Found Riddled with Leaks," *Boston Globe*, Boston.com, Nov. 10, 2004; William J. Angelo,

"Concrete Supplier Fined in Central Artery/Tunnel Scam," *Engineering News-Record,* Aug. 6, 2007, p. 13.

7. Angelo, "Concrete Supplier Fined in Central Artery/Tunnel Scam."

8. Matthew L. Wald, "Late Design Change Is Cited in Collapse of Tunnel Ceiling," *New York Times,* Nov. 2, 2006, p. A18.

9. "Bits and Building Work May Be Factors in CA/T Collapse," *Engineering News-Record,* Aug. 7, 2006, p. 16.

10. William J. Angelo, "Collapse Report Stirs Debate on Epoxies," *Engineering News-Record,* July 23, 2007, pp. 10–12; William J. Angelo, "Epoxy Supplier Challenges Boston Tunnel Report," *Engineering News-Record,* July 26, 2007, http://enr.construction.com/news/transportation/archives/070726.asp.

11. Ken Belson, "A Mix of Sand, Gravel and Glue That Drives the City Ever Higher," *New York Times,* June 21, 2008, http://www.nytimes.com/2008/06/21/nyregion/21industry.html.

12. For the Salginatobel Bridge, see, e.g., David P. Billington, *Robert Maillart's Bridges: The Art of Engineering* (Princeton, N.J.: Princeton University Press, 1979), chap. 8; Galinsky, "TWA Terminal, John F. Kennedy Airport NY," http://www.galinsky.com/buildings/twa/index.htm; "Dubai to Open World's Tallest Building," Breitbart.com, Jan. 1, 2010.

13. Tony Illia, "Poor, Often-Homemade Concrete Blamed for Much Haiti Damage," *Engineering News-Record,* Feb. 4, 2010, http://enr.ecnext.com/coms2/article_bucm100204HaitiPoorCon; Associated Press, "Haiti Bans Construction Using Quarry Sand," Feb. 14, 2010; Ayesha Bhatty, "Haiti Devastation Exposes Shoddy Construction," *BBC News,* Jan. 15, 2010, http://news.bbc.co.uk/2/hi/americas/8460042.stm; Tom Sawyer and Nadine Post, "Haiti's Quake Assessment Is Small Step Toward Recovery," *Engineering News-Record,* Feb. 1, 2010, pp. 12–13.

14. Reginald DesRoches, Ozlem Ergun, and Julie Swann, "Haiti's Eternal Weight," *New York Times,* July 8, 2010, p. A25; Nadine M. Post, "Engineers Fear Substandard Rebuilding Coming in Quake-Torn Haiti," ENR.com, Feb. 3, 2010, http://enr.ecnext.com/coms2/article_inen100203QuakeTornHai; Associated Press, "Haiti Bans Construction Using Quarry Sand."

15. "Building Collapse Kills One Worker in Shanghai," *China Daily,* June 27, 2009, http://www.chinadaily.com.cn/china/2009–06/27/content_833067.htm; slideshow attachment to e-mail from Bruce Kirstein to author, Jan. 5, 2010; Associated Press, "Building, Factory Wall in China Topple, Killing 14," Oct. 3, 2010, http://enr.construction.com/yb/enr/article.aspx?story_id=150572484;   Agence France-Presse, "Building Collapse Kills Eight in China," Oct. 3, 2010, http://www.google.com/hostednews/afp/article/ALeqM5h3PvA_fSfvuYdDSv3gyTRGl6DwEw?docId=CNG.23111bf2d9c2a75f1ce1e14c0bcb1919.261; "Sichuan   Earthquake,

Poorly-Built Schools and Parents: Schools Hit by the Sichuan Earthquake in 2008," *FactsandDetails.com*, http://factsanddetails.com/china.php?itemid=1020 &catid=10&subcatid=65.

16. Kirstein to author, Jan. 5, 2010.

17. Post, "Engineers Fear Substandard Rebuilding Coming"; William K. Rashbaum, "Company Hired to Test Concrete Faces Scrutiny," *New York Times*, June 21, 2008, http://www.nytimes.com/2008/06/21/nyregion/21concrete.html.

18. Richard Korman, "Indictment Filed against New York's Biggest Concrete Testing Laboratory," *Engineering News-Record*, Oct. 30, 2008, http://enr.ecnext. com/coms2/article_nefiar081030; John Eligon, "Concrete-Testing Firm Is Accused of Skipping Tests," *New York Times*, Oct. 31, 2008, p. A27.

19. Metropolitan Transportation Authority, "Second Avenue Subway," http:// www.mta.info/capconstr/sas; Korman, "Indictment Filed"; Colin Moynihan, "Concrete Testing Executive Sentenced to up to 21 Years," *New York Times*, May 17, 2010, p. A28.

20. Sushil Cheema, "The Big Dig: The Yanks Uncover a Red Sox Jersey," *New York Times*, April 14, 2008, http://www.nytimes.com/2008/04/14/sports/ baseball/14jersey.html; William K. Rashbaum and Ken Belson, "Cracks Emerge in Ramps at New Yankee Stadium," *New York Times*, Oct. 24, 2009, http://www. nytimes.com/2009/10/24/nyregion/24stadium.html; see also William K. Rashbaum, "Concrete Testing Inquiry Widens to Include a Supplier for Road Projects," *New York Times*, Aug. 18, 2009, p. A17.

21. David McCullough, *The Great Bridge* (New York: Simon & Schuster, 1972), pp. 374, 444–445.

22. Ibid., pp. 442–447.

23. Leslie Wayne, "Thousands of Homeowners Cite Drywall for Ills," *New York Times*, Oct. 8, 2009, http://www.nytimes.com/2009/10/08/business/08drywall. html; Julie Schmit, "Drywall from China Blamed for Problems in Homes," *USA Today*, March 17, 2009, http://www.usatoday.com/money/economy/housing/ 2009–03–16-chinese-drywall-sulfur_N.htm; Pam Hunter, Scott Judy, and Sam Barnes, "Paying to Replace Chinese Drywall," *Engineering-News Record*, April 19, 2010, pp. 10–11; Andrew Martin, "Drywall Flaws: Owners Gain Limited Relief," *New York Times*, Sept. 18, 2010, pp. A1, A3.

24. Mary Williams Walsh, "Bursting Pipes Lead to a Legal Battle," *New York Times*, Feb. 12, 2010, http://www.nytimes.com/2010/02/12/business/12pipes.html; Tom Sawyer, "PVC Pipe Firm's False-Claims Suit Unsealed by District Court," *Engineering News-Record*, Feb. 22, 2010, p. 15; Mary Williams Walsh, "Facing Suit, Pipe Maker Extends Guarantee," *New York Times*, April 6, 2010, pp. B1, B2.

25. Walsh, "Bursting Pipes Lead to a Legal Battle."

26. David Crawford, Reed Albergotti, and Ian Johnson, "Speed and Commerce Skewed Track's Design," *Wall Street Journal*, Feb. 16, 2010.

27. Ibid.; John Branch and Jonathan Abrams, "Luge Athlete's Death Casts Pall over Olympics," *New York Times*, Feb. 13, 2010, graphic, p. D2.

28. Crawford, Albergotti, and Johnson, "Speed and Commerce Skewed Track's Design."

29. David Crawford and Matt Futterman, "Luge Track Had Earlier Fixes Aimed at Safety," *Wall Street Journal*, Feb. 20, 2010, http://online.wsj.com/article/SB10001424052748703787304575075383263999728.html.

30. Branch and Abrams, "Luge Athlete's Death Casts Pall over Olympics," pp. A1, D2.

31. Crawford, Albergotti, and Johnson, "Speed and Commerce Skewed Track's Design."

2. Things Happen

1. Diane Vaughan, *The Challenger Launch Decision: Risky Technology, Culture, and Deviance at NASA* (Chicago: University of Chicago Press, 1996), p. 274; R. P. Feynman, "Personal Observations of the Reliability of the Shuttle," see http://www.ralentz.com/old/space/feynman-report.html.

2. John Schwartz, "Minority Report Faults NASA as Compromising Safety," *New York Times*, Aug. 18, 2005, p. A18.

3. Todd Halvorson, "'We Were Lucky': NASA Underestimated Shuttle Dangers," *Florida Today*, Feb. 13, 2011, http://floridatoday.com/article/20110213/NEWS02/102130319/1007/NEWS02/We-were-lucky-NASA-underestimated-shuttle-dangers.

4. Newton quoted in John Bartlett, *Familiar Quotations*, 16th ed., Justin Kaplan, gen. ed. (Boston: Little, Brown, 1992), p. 303.

5. Newton quoted in Bartlett, *Familiar Quotations*, 16th ed., p. 281.

6. Gary Brierley, "Free to Fail," *Engineering News-Record*, April 25, 2011, p. U6.

7. Mark Schrope, "The Lost Legacy of the Last Great Oil Spill," *Nature*, July 15, 2010, pp. 304–305; Jo Tuckman, "Gulf Oil Spill: Parallels with Ixtoc Raise Fears of Ecological Tipping Point," *Guardian.co.uk*, June 1, 2010, http://www.guardian.co.uk/environment/2010/jun/01/gulf-oil-spill-ixtoc-ecological-tipping-point; Edward Tenner, "Technology's Disaster Clock," *The Atlantic*, June 18, 2010, http://www.theatlantic.com/science/archive/2010/06/technologys-disaster-clock/58367.

8. On bridge failures, see Paul Sibly, "The Prediction of Structural Failures," Ph.D. thesis. University of London, 1977.

9. Riddle of the Sphinx quoted in Bartlett, *Familiar Quotations,* 16th ed., p. 66, n. 1.

10. Vitruvius, *The Ten Books on Architecture,* trans. Morris Hicky Morgan (New York: Dover Publications, 1960), p. 80 (III, III, 4); compare Henry Petroski, *Design Paradigms: Case Histories of Error and Judgment in Engineering* (New York: Cambridge University Press, 1994), chaps. 2, 3; Henry Petroski, "Rereading Vitruvius," *American Scientist,* Nov.–Dec. 2010, pp. 457–461.

11. Vitruvius, *Ten Books on Architecture,* pp. 285–289 (X, II, 1–14).

12. Galileo, *Dialogues Concerning Two New Sciences,* H. Crew and A. de Salvio, trans. (New York: Dover Publications, [1954]), pp. 2, 4–5, 131.

13. Ibid., pp. 115–118; Petroski, *Design Paradigms,* pp. 64–74.

14. Galileo, *Dialogues Concerning Two New Sciences,* p. 5.

15. See Petroski, *Design Paradigms,* chap. 4. For background on the Hyatt Regency failure see Henry Petroski, *To Engineer Is Human: The Role of Failure in Successful Design* (New York: St. Martin's Press, 1985), chap. 8.

16. Joe Morgenstern, "The Fifty-Nine Story Crisis," *New Yorker,* May 29, 1995, pp. 45–53.

17. Ibid.

18. See, e.g., Linda Geppert, "Biology 101 on the Internet: Dissecting the Pentium Bug," *IEEE Spectrum,* Feb. 1996, pp. 16–17.

19. Miguel Helft, "Apple Confesses to Flaw in iPhone's Signal Meter," *New York Times,* July 3, 2010, pp. B1, B2; Apple, "Letter from Apple Regarding iPhone 4," July 2, 2010, http://www.apple.com/pr/library/2010/07/02appleletter.html; Matthias Gross to author, letter dated June 27, 2011; see also Matthias Gross, *Ignorance and Surprise: Science, Society, and Ecological Design* (Cambridge, Mass.: MIT Press, 2010).

20. Mike Gikas, "Lab Tests: Why Consumer Reports Can't Recommend the iPhone 4," Electronics Blog, *ConsumerReports.com,* July 13, 2010, http://blogs.consumerreports.org/electronics/2010/07/apple-iphone-4-antenna-issue-iphone4-problems-dropped-calls-lab-test-confirmed-problem-issues-signal-strength-att-network-gsm.html; "Apple iPhone 4 Bumper—Black," http://store.apple.com/us/product/MC597ZM/A#overview.

21. Peter Burrows and Connie Guglielmo, "Apple Engineer Told Jobs IPhone Antenna Might Cut Calls," *Bloomberg,* July 15, 2010, http://www.bloomberg.com/news/2010–07–15/apple-engineer-said-to-have-told-jobs-last-year-about-iphone-antenna-flaw.html; Bloomberg Business Week, "Apple Sets Up Cots for Engineers Solving iPhone Flaw," *Bloomberg.com,* July 17, 2010, http://www.businessweek.com/news/2010–07–17/apple-sets-up-cots-for-engineers-solving-iphone-flaw.html. Further iPhone glitches included alarms not going off on New

Year's Day and clocks falling back instead of springing forward one hour when daylight savings time went into effect in 2011: Associate Press, "Some iPhones Bungle Time Change," *Herald-Sun* (Durham, N.C.), March 14, 2011, p. A4.

22. Miguel Helft and Nick Bilton, "Design Flaw in iPhone 4, Testers Say," *New York Times*, July 13, 2010, http://www.nytimes.com/2010/07/13/technology/13apple. html?_r=1&emc=eta1; Gikas, "Lab Tests: Why Consumer Reports Can't Recommend the iPhone 4"; Gross, *Ignorance and Surprise*, p. 32.

23. Newton quoted in Bartlett, *Familiar Quotations*, 16th ed., p. 281.

3. Designed to Fail

1. Russ McQuaid, "Piece Reattached after Coming Loose in Moderate Wind before Fatal Concert," *Fox59.com*, Aug. 18, 2011, http://www.fox59.com/news/wxin-grandstand-collapse-investigates-roof-fox59-investigates-condition-of-grandstands-roof-before-collapse-20110818,0,2486205.column.

2. For a description of the Apollo 13 accident, see, e.g., Charles Perrow, *Normal Accidents: Living with High-Risk Technologies* (Princeton, N.J.: Princeton University Press, 1999), pp. 271–278.

3. Steven J. Paley, *The Art of Invention: The Creative Process of Discovery and Design* (Auburn, N.Y.: Prometheus Books, 2010), pp. 157–159.

4. NACE International, "1988—The Aloha Incident," http://events.nace.org/library/corrosion/aircraft/aloha.asp; Wikipedia, "Aloha Airlines Flight 243," http://en.wikipedia.org/wiki/Aloha_Airlines_Flight_243.

5. Christopher Drew and Jad Mouawad, "Boeing Says Jet Cracks Are Early," *New York Times*, April 6, 2011, pp. B1, B7; Matthew L. Wald and Jad Mouawad, "Rivet Flaw Suspected in Jet's Roof," *New York Times*, April 26, 2011, pp. B1, B4.

6. Rainer F. Foelix, *Biology of Spiders* (Cambridge, Mass.: Harvard University Press, 1982), pp. 146–147.

7. On zippers, see, e.g., Henry Petroski, *The Evolution of Useful Things* (New York: Alfred A. Knopf, 1992), chap. 6, and Henry Petroski, *Invention by Design: How Engineers Get from Thought to Thing* (Cambridge, Mass.: Harvard University Press, 1996), chap. 4; see also Robert Friedel, *Zipper: An Exploration in Novelty* (New York: W. W. Norton, 1994).

8. Brett Stern, *99 Ways to Open a Beer Bottle without a Bottle Opener* (New York: Crown, 1993); Will Gottlieb, "Warning: 'Twist off' Means Twist Off," (Maine) *Coastal Journal*, June 23, 2011, p. 36.

9. On pop-top cans, see Petroski, *Evolution of Useful Things*, chap. 11.

10. Caltrans, "The San Francisco—Oakland Bay Bridge Seismic Safety Projects," http://baybridgeinfo.org/seismic_innovations.

11. Bureau of Reclamation, *Reclamation: Managing Water in the West* (U.S. Department of the Interior, 2006), p. 20.

12. Randy Kennedy, "Yankee Stadium Closed as Beam Falls onto Seats," *New York Times*, April 14, 1998, pp. A1, C3.

13. Douglas A. Anderson, "The Kingdome Implosion," *Journal of Explosives Engineering* 17, no. 5 (Sept.–Oct. 2000), pp. 6–15.

14. South Carolina Department of Transportation, "Cooper River Bridge: History," http://www.cooperriverbridge.org/history.shtml; Charles Dwyer, "Cooper River Bridge Demolition," annotated slide presentation, 36th Annual South Carolina State Highway Conference, Clemson, S.C., March 28, 2007, http://www.clemson.edu/t3s/workshop/2007/Dwyer.pdf.

15. Dwyer, "Cooper River Bridge Demolition," slides 13–87.

16. New York State Department of Transportation, "Lake Champlain Bridge Project," https://www.nysdot.gov/lakechamplainbridge/history; see also Christopher Kavars, "The Nuts and Bolts of Dynamic Monitoring," *Structural Engineering and Design*, Dec. 2010, pp. 26–29.

17. Aileen Cho, "Officials Hurrying with Plans to Replace Closed Crossing," *Engineering News-Record*, November 20, 2009, p. 15; Lohr McKinstry, "Flatiron Wins Contract for New Champlain Bridge," *ENR.com*, May 29, 2010, http://enr.construction.com/yb/enr/article.aspx?story_id=145562812&elq=182e4fdeabe54cb69a1945be86b3afc7.

18. "Controlled Explosives Topple Aging Champlain Bridge," *WPTZ.com*, Dec. 21, 2009, http://www.wptz.com/print/22026547/detail.html.

19. See, e.g., Donald Simanek, "Physics Lecture Demonstrations, with Some Problems and Puzzles, Too," http://www.lhup.edu/~dsimanek/scenario/demos.htm, s.v. "Chimney Toppling"; Gabriele Varieschi, Kaoru Kamiya, and Isabel Jully, "The Falling Chimney Web Page," http://myweb.lmu.edu/gvarieschi/chimney/chimney.html.

20. Mike Larson, "Tower Knockdown Scheme Does Not Go as Planned," *Engineering News-Record*, Nov. 22, 2010, p. 14.

21. Nadine M. Post, "Faulty Tower's Implosion Will Set New Record," *Engineering News-Record*, Nov. 30, 2009, pp. 12–13.

22. Ibid.

23. David Wolman, "Turning the Tides," *Wired*, Jan. 2009, pp. 109–113.

24. Ibid.

25. The Encyclopedia of Earth, "Price-Anderson Act of 1957, United States," http://www.eoearth.org/article/Price-Anderson_Act_of_1957,_United_States.

26. Barry B. LePatner, *Too Big to Fall: America's Failing Infrastructure and the Way Forward* (New York: Foster Publishing, 2010), p. 173.

4. Mechanics of Failure

1. Department of Theoretical and Applied Mechanics, *The Times of TAM,* brochure (Urbana, Ill.: TAM Department, [2006]).

2. Ernst Mach, *The Science of Mechanics: A Critical and Historical Account of Its Development,* Thomas J. McCormack, trans. (La Salle, Ill.: Open Court, 1960), p. 1.

3. Instron, "History of SATEC," www.instron.com/wa/library/StreamFile.aspx?doc=466; Shakhzod M. Takhirov, Dick Parsons, and Don Clyde, "Documentation of the 4 Million Pound Southwark-Emery Universal Testing Machine," Earthquake Engineering Research Center, University of California, Berkeley, Aug. 2004, http://nees.berkeley.edu/Facilities/pdf/4MlbsUTM/4Mlb_Southwark_Emery_UTM.pdf.

4. James W. Phillips to author, e-mail message, June 14, 2010.

5. William Rosen, *The Most Powerful Idea in the World: A Story of Steam, Industry, and Invention* (New York: Random House, 2010), p. 68; Mark Gumz, quoted in Harold Evans, Gail Buckland, and David Lefer, *They Made America: From the Steam Engine to the Search Engine: Two Centuries of Innovators* (New York: Little, Brown, 2004), p. 465; Wen Hwee Liew to author, e-mail message dated July 19, 2011.

6. Timothy P. Dolen, "Advances in Mass Concrete Technology—The Hoover Dam Studies," in *Proceedings, Hoover Dam 75th Anniversary History Symposium,* Richard L. Wiltshire, David R. Gilbert, and Jerry R. Rogers, eds., American Society of Civil Engineers Annual Meeting, Las Vegas, Nev., Oct. 21–22, 2010, pp. 58–73; Katie Bartojay and Westin Joy, "Long-Term Properties of Hoover Dam Mass Concrete," ibid., pp. 74–84.

7. Ron Landgraf, ed., *JoMo Remembered: A Tribute Volume Celebrating the Life and Career of JoDean Morrow, Teacher, Researcher, Mentor and International Bon Vivant* (privately printed, 2009), p. 2.

8. Morrow quoted in ibid., p. 21.

9. James W. Phillips, compiler and editor, *Celebrating TAM's First 100 Years: A History of the Department of Theoretical and Applied Mechanics, University of Illinois at Urbana-Champaign, 1890–1990* (Urbana, Ill.: TAM Department, 1990), p. PHD-6.

10. Quoted in Rosen, *The Most Powerful Idea in the World,* p. 67.

11. For a broad perspective on fracture, see Brian Cotterell, *Fracture and Life* (London: Imperial College Press, 2010).

12. Stanley T. Rolfe and John M. Barsom, *Fracture and Fatigue Control in*

Structures: Applications of Fracture Mechanics (Englewood Cliffs, N.J.: Prentice-Hall, 1977).

13. American Society for Testing and Materials, *ASTM 1898–1998: A Century of Progress* (West Conshohocken, Pa.: ASTM, 1998). It was for ASTM's Committee E08 that I prepared the 2006 Annual Fatigue Lecture that grew into this chapter and the next.

14. Henry Petroski, *To Engineer Is Human: The Role of Failure in Successful Design* (New York: St. Martin's Press, 1985); Henry Petroski, "On the Fracture of Pencil Points," *Journal of Applied Mechanics* 54 (1987): 730–733; Henry Petroski, *The Pencil: A History of Design and Circumstance* (New York: Knopf, 1990); Henry Petroski, *The Evolution of Useful Things* (New York: Knopf, 1992).

5. A Repeating Problem

1. Jason Annan and Pamela Gabriel, *The Great Cooper River Bridge* (Columbia: University of South Carolina Press, 2002).

2. Peter R. Lewis, "Safety First?" *Mechanical Engineering,* Sept. 2010, pp. 32–35.

3. Galileo, *Dialogues Concerning Two New Sciences,* trans. H. Crew and A. de Salvio (New York: Dover Publications, [1954]), pp. 2, 5.

4. William John Macquorn Rankine, "On the Causes of the Unexpected Breakage of the Journals of Railway Axles; and on the Means of Preventing Such Accidents by Observing the Law of Continuity in Their Construction," *Minutes of the Proceedings of the Institution of Civil Engineers* 2 (1843): 105–108.

5. Peter R. Lewis, *Disaster on the Dee: Robert Stephenson's Nemesis of 1847* (Stroud, Gloucestershire: Tempus Publishing, 2007); "Report of the Commissioners Appointed to Inquire into the Application of Iron to Railway Structures," *Journal of the Franklin Institute,* June 1850, p. 365; Derrick Beckett, *Stephensons' Britain* (Newton Abbot, Devon: David & Charles, 1984), pp. 123–125.

6. Lewis, *Disaster on the Dee,* pp. 93, 113–114, 121.

7. Ibid., pp. 95–109.

8. Ibid., pp. 116–118, 138–139.

9. Ibid., pp. 95, 103–104.

10. Peter Rhys Lewis, Ken Reynolds, and Colin Gagg, *Forensic Materials Engineering: Case Studies* (Boca Raton, Fla.: CRC Press, 2004); Peter R. Lewis and Colin Gagg, "Aesthetics versus Function: The Fall of the Dee Bridge, 1847," *Interdisciplinary Science Reviews* 29 (2004), 2: 177–191.

11. Lewis and Gagg, "Aesthetics versus Function."

12. See, e.g., Henry Petroski, *Design Paradigms: Case Histories of Error and Judgment in Engineering* (New York: Cambridge University Press, 1994), pp. 83–84.

13. Wikipedia, "Liverpool and Manchester Railway," http://en.wikipedia.org/wiki/Liverpool_and_Manchester_Railway; Peter R. Lewis and Alistair Nisbet, *Wheels to Disaster! The Oxford Train Wreck of Christmas Eve 1874* (Stroud, Gloucestershire: Tempus, 2008), pp. 56–76.

14. Frederick Braithwaite, "On the Fatigue and Consequent Fracture of Metals," *Minutes of the Proceedings of the Institution of Civil Engineers* 13 (1854): 463–467. See also discussion, ibid., pp. 467–475; Poncelet quoted in J. Y. Mann, "The Historical Development of Research on the Fracture of Materials and Structures," *Journal of the Australian Institute of Metals* 3 (1958), 3: 223. See also Walter Schütz, "A History of Fatigue," *Engineering Fracture Mechanics* 54 (1996), 2: 263–300.

15. Peter R. Lewis and Ken Reynolds, "Forensic Engineering: A Reappraisal of the Tay Bridge Disaster," *Interdisciplinary Science Reviews* 27 (2002), 4: 287–298; see also T. Martin and I. MacLeod, "The Tay Bridge Disaster: A Reappraisal Based on Modern Analysis Methods," *Proceedings of the Institution of Civil Engineers* 108 (1995), Civil Engineering: 77–83. For the story of the Tay Bridge, see also Peter R. Lewis, *Beautiful Railway Bridge of the Silvery Tay: Reinvestigating the Tay Bridge Disaster of 1879* (Stroud, Gloucestershire: Tempus, 2004).

16. Lewis and Reynolds, "Forensic Engineering," pp. 288, 290.

17. Lewis, *Beautiful Railway Bridge*, p. 69.

18. Ibid., p. 129.

19. Ibid., pp. 74, 75–76.

20. Ibid., pp. 70, 90; Lewis and Reynolds, "Forensic Engineering," p. 288.

21. Lewis, *Beautiful Railway Bridge*, pp. 133–134; Peter Lewis to author, e-mail message, September 22, 2010.

22. Lewis, *Beautiful Railway Bridge*, pp. 134–148.

23. B. Baker, *Long-Span Railway Bridges*, rev. ed. (London: Spon, 1873), p. 90; R. A. Smith, "The Wheel-Rail Interface—Some Recent Accidents," *Fatigue and Fracture in Engineering Materials and Structures* 26 (2003): 901–907; Matthew L. Wald, "Seaplane Fleet to Be Tested for Metal Fatigue after Crash," *New York Times*, Dec. 23, 2005, p. A16. See also R. A. Smith, "Railway Fatigue Failures: An Overview of a Long-Standing Problem," *Materialwissenschaft und Werkstofftechnik* 36 (2005): 697–705.

24. Sante Camo, "The Evolution of a Design," *Structural Engineer*, Jan. 2004, pp. 32–37.

25. Michael Cabanatuan, "Bay Bridge Officials Plan to Prevent Cracks," *San Francisco Chronicle*, July 15, 2010, http://www.sfgate.com/cgi-bin/article.cgi?f=/

c/a/2010/07/15/BAMU1EEJ8D.DTL; "Crews Find Bay Bridge Is Cracked," *New York Times*, Sept. 7, 2009, p. A10; "San Francisco Artery Reopens after Second Emergency Fix," *Engineering News-Record*, Nov. 9, 2009, p. 52.

6. The Old and the New

1. *Bridges—Technology and Insurance* (Munich: Munich Reinsurance Company, 1992), p. 85.

2. D. B. Steinman and C. H. Gronquist, "Building First Long-Span Suspension Bridge in Maine," *Engineering News-Record*, March 17, 1932, pp. 386–389.

3. Australian Transport Safety Bureau, "In-Flight Uncontained Engine Failure Overhead Batam Island, Indonesia, 4 November 2010," ATSB Transport Safety Report, Aviation Occurrence Investigation AO-2010–089, Preliminary (Canberra City: ATSB, 2010).

4. Stephen C. Foster, *Building the Penobscot Narrows Bridge and Observatory*, photographic essay ([Woolwich, Maine]: Cianbro/Reed & Reed, 2007), p. [vi]. Beginning in 2010, trucks weighing up to 100,000 pounds were allowed on the interstate highways in Maine and Vermont, thus making it less likely that they would have to use non-interstate roads like U.S. 1. See David Sharp, "Maine, Vermont Hail Truck-Weight Exemption," Burlingtonfreepress.com, Jan. 11, 2010; see also William B. Cassidy, "White House Backs Bigger Trucks in Maine, Vermont," *Journal of Commerce Online*, Sept. 20, 2010, http://www.joc.com/trucking/white-house-backs-bigger-trucks-maine-vermont.

5. Steinman and Gronquist, "Building First Long-Span Suspension Bridge in Maine," p. 386.

6. Peter Taber, "DOT Makes Urgent Call for New Bridge," *Waldo* (Maine) *Independent*, July 3, 2003, pp. 1, 9.

7. Bill Trotter, "Bridge Truck Ban Raises Anxiety," *Bangor* (Maine) *Daily News*, July 14, 2003, pp. A1, A8; William J. Angelo, "Maine Cables Get Extra Support in Rare Procedure," *Engineering News-Record*, Nov. 10, 2003, pp. 24, 27.

8. Foster, *Building the Penobscot Narrows Bridge and Observatory*, p. 48.

9. Ibid., p. 132; *Penobscot Narrows Bridge and Observatory*, official commemorative brochure, May 2007, pp. 12–13; Eugene C. Figg, Jr., and W. Denney Pate, "Cable-Stay Cradle System," U.S. Patent No. 7,003,835 (Feb. 28, 2006).

10. *Penobscot Narrows Bridge and Observatory*, p. 10; Foster, *Building the Penobscot Narrows Bridge and Observatory*, p. 56.

11. "And the Winner Is: Downeast Gateway Bridge," *Boston Globe*, Jan. 8, 2006, http://www.boston.com/news/local/maine/articles/2006/01/08/and_the_winner_is_downeast_gateway_bridge.

12. Richard G. Weingardt, *Circles in the Sky: The Life and Times of George Ferris* (Reston, Va.: ASCE Press, 2009), p. 88.

13. Foster, *Building the Penobscot Narrows Bridge and Observatory,* p. 57.

14. "The End of an Era . . . and the Beginning of Another," *Bucksport* (Maine) *Enterprise,* Jan. 4, 2007, p. 1.

15. Rich Hewitt, "Old Span's Removal Not Expected Soon," *Bangor* (Maine) *Daily News,* Oct. 14–15, 2006, p. A7; Larry Parks to author, e-mail message, Aug. 23, 2007.

16. Parks to author.

7. Searching for a Cause

1. See, e.g., Virginia Kent Dorris, "Hyatt Regency Hotel Walkways Collapse," in *When Technology Fails: Significant Technological Disasters, Accidents, and Failures of the Twentieth Century,* ed. Neil Schlager (Detroit: Gale Research, 1994), pp. 317–325; Norbert J. Delatte, Jr., *Beyond Failure: Forensic Case Studies for Civil Engineers* (Reston, Va.: ASCE Press, 2009), pp. 8–25; Henry Petroski, *To Engineer Is Human: The Role of Failure in Successful Design* (New York: St. Martin's Press, 1985), chap. 8 and illustration section following p. 106.

2. Rita Robison, "Point Pleasant Bridge Collapse," in *When Technology Fails,* p. 202.

3. Ibid., p. 203.

4. Delatte, *Beyond Failure,* p. 71; Robison, "Point Pleasant Bridge Collapse," pp. 202, 204; Abba G. Lichtenstein, "The Silver Bridge Collapse Recounted," *Journal of Performance of Constructed Facilities* 7, 4 (Nov. 1993): 251, 255–256.

5. Delatte, *Beyond Failure,* pp. 71–73; Robison, "Point Pleasant Bridge Collapse," p. 203; Wilson T. Ballard, "An Eyebar Suspension Span for the Ohio River," *Engineering News-Record,* June 20, 1929, p. 997.

6. Lichtenstein, "The Silver Bridge Collapse Recounted," p. 256; "Bridge Failure Probe Shuts Twin," *Engineering News-Record,* Jan. 9, 1969, p. 17.

7. Corrosion Doctors, "Silver Bridge Collapse," http://www.corrosion-doctors.org/Bridges/Silver-Bridge.htm.

8. "Cause of Silver Bridge Collapse Studied," *Civil Engineering,* Dec. 1968, p. 87; Robison, "Point Pleasant Bridge Collapse," pp. 203, 205; Lichtenstein, "Silver Bridge Collapse Recounted," p. 259; Robert T. Ratay, "Changes in Codes, Standards and Practices Following Structural Failures, Part 1: Bridges," *Structure,* Dec. 2010, p. 16; "Rules and Regulations," *Federal Register* 69, 239 (Dec. 14, 2004), p. 74,419.

9. "Collapsed Silver Bridge Is Reassembled," *Engineering News-Record,* April

25, 1968, pp. 28–30; Charles F. Scheffey, "Pt. Pleasant Bridge Collapse: Conclusions of the Federal Study," *Civil Engineering*, July 1971, pp. 41–45.

10. Delatte, *Beyond Failure*, p. 74; "Collapsed Silver Bridge Is Reassembled," p. 29.

11. "Bridge Failure Triggers Rash of Studies," *Engineering News-Record*, Jan. 4, 1968, p. 18.

12. Delatte, *Beyond Failure*, p. 75; Scheffey, "Pt. Pleasant Bridge Collapse," p. 42.

13. Scheffey, "Pt. Pleasant Bridge Collapse," p. 42.

14. W. Jack Cunningham to author, letter dated April 24, 1995; S. Reier, *The Bridges of New York* (New York: Quadrant Press, 1977), p. 47; "Birds on Big Bridge Vouch Its Strength," *New York Times*, Dec. 10, 1908, p. 3; Edward E. Sinclair, "Birds on New Bridge," letter to the editor, *New York Times*, Dec. 1, 1908. The reference to Kipling's "The Bridge Builders" is erroneous; perhaps the story of birds and bridges appears in another of his stories.

15. Chris LeRose, "The Collapse of the Silver Bridge," *West Virginia Historical Society Quarterly* XV (2001), http://www.wvculture.org/history/wvhs1504.html; Infoplease, "Chief Cornstalk," http://www.infoplease.com/ipa/A0900079.html.

16. Robison, "Point Pleasant Bridge Collapse," p. 202; Delatte, *Beyond Failure*, p. 74.

17. Lichtenstein, "Silver Bridge Collapse Recounted," pp. 249, 253–254; Robison, "Point Pleasant Bridge Collapse," p. 204.

18. Delatte, *Beyond Failure*, p. 75; Robison, "Point Pleasant Bridge Collapse," p. 204.

19. Daniel Dicker, "Point Pleasant Bridge Collapse Mechanism Analyzed," *Civil Engineering*, July 1971, pp. 61–66.

20. "Cause of Silver Bridge Collapse Studied"; Dicker, "Point Pleasant Bridge Collapse Mechanism," p. 64; Lichtenstein, "Silver Bridge Collapse Recounted," p. 260.

21. Stanley T. Rolfe and John M. Barsom, *Fracture and Fatigue Control in Structures: Applications of Fracture Mechanics* (Englewood Cliffs, N.J.: Prentice-Hall, 1977), pp. 2–4.

22. Delatte, *Beyond Failure*, p. 77; Rolf and Barsom, *Fracture and Fatigue Control*, pp. 13, 22. The distinction between stress-corrosion and corrosion-fatigue cracking is clarified in Joe Fineman to the editors of *American Scientist*, e-mail message dated Aug. 16, 2011.

23. National Transportation Safety Board, *Highway Accident Report: Collapse of U.S. 35 Highway Bridge, Point Pleasant, West Virginia, December 15, 1967*, Report No. NTSB-HAR-71–1, p. 126.

24. Robison, "Point Pleasant Bridge Collapse," p. 202.

25. Associated Press, "10 Years After TWA 800, Doubts Abound," *msnbc.com,* July 8, 2006, http://www.msnbc.msn.com/id/13773369.

26. Ibid.

27. William J. Broad, "Hard-Pressed Titanic Builder Skimped on Rivets, Book Says," *New York Times,* April 15, 2008, p. A1.

28. Ibid.

29. Ibid.

30. "Great Lakes' Biggest Ship to Be Launched Tomorrow," *New York Times,* June 6, 1958, p. 46.

31. Ibid.; "The Sinking of the SS Edmund Fitzgerald—November 10, 1975," http://cimss.ssec.wisc.edu/wxwise/fitz.html; Great Lakes Shipwreck Museum, "Edmund Fitzgerald," http://www.shipwreckmuseum.com/edmundfitzgerald.

8. The Obligation of an Engineer

1. For the story of the Quebec Bridge, see, e.g., William D. Middleton, *The Bridge at Quebec* (Bloomington: Indiana University Press, 2001); see also Henry Petroski, *Engineers of Dreams: Great Bridge Builders and the Spanning of America* (New York: Alfred A. Knopf, 1995), pp. 101–118.

2. Yale University, "History of Yale Engineering," http://www.seas.yale.edu/about-history.php; "Yale Engineering through the Centuries," http://www.eng.yale.edu/eng150/timeline/index.html; American Physical Society, "J. Willard Gibbs," http://www.aps.org/programs/outreach/history/historicsites/gibbs.cfm.

3. Bruce Fellman, "The Rebuilding of Engineering," *Yale Magazine,* Nov. 1994, pp. 36–41.

4. *Who's Who in America,* 1994; Fellman, "Rebuilding of Engineering."

5. Fellman, "Rebuilding of Engineering," pp. 37, 39.

6. *Who's Who in America;* Fellman, "Rebuilding of Engineering," p. 39.

7. National Institutes of Health, "The Hippocratic Oath," http://www.nlm.nih.gov/hmd/greek/greek_oath.html; American Society of Civil Engineers, *Official Register* (Reston, Va.: ASCE, 2009), p. 13.

8. Fellman, "Rebuilding of Engineering," p. 39; Donald H. Jamieson, "The Iron Ring—Myth and Fact," unsourced photocopy; Wikipedia, "Iron Ring," http://en.wikipedia.org/wiki/Iron_Ring.

9. Norman R. Ball, "The Iron Ring: An Historical Perspective," *Engineering Dimensions,* March/April 1991, pp. 46, 48.

10. Peter R. Hart, *A Civil Society: A Brief Personal History of the Canadian So-*

ciety for Civil Engineering (Montreal: Canadian Society for Civil Engineering, 1997).

11. Jamieson, "The Iron Ring—Myth and Fact"; see also Augustine J. Fredrich, ed., *Sons of Martha: Civil Engineering Readings in Modern Literature* (New York: American Society of Civil Engineers, 1989), p. 63.

12. See Ball, "The Iron Ring," p. 48; see also John A. Ross, "Social Specifications for the Iron Ring," *BC Professional Engineer*, Aug. 1980, pp. 12–18; Fekri S. Osman, "The Iron Ring," *IEEE Engineering in Medicine and Biology Magazine*, June 1984, p. 39; Don Shields to author, letter dated April 14, 1998.

13. Bill Bryson, ed., *Seeing Further: The Story of the Royal Society* (London: HarperPress, 2010), endpapers.

14. "Ritual of the Calling of an Engineer, 1925–2000," *Canada's Stamp Details* 9, no. 2 (March/April 2000), http://www.canadapost.ca/cpo/mc/personal/collecting/stamps/archives/2000/2000_apr_ritual.jsf; "The Iron Ring: The Ritual of the Calling of an Engineer/Les rites d'engagement de l'ingenieur," http://www.ironring.ca; Ball, "The Iron Ring"; Osman, "The Iron Ring"; Rudyard Kipling, "Cold Iron," *PoemHunter.com*, http://www.poemhunter.com/poem/cold-iron. See also Robin S. Harris and Ian Montagnes, eds., *Cold Iron and Lady Godiva: Engineering Education at Toronto, 1920–1972* (Toronto: University of Toronto Press, 1973).

15. Jamieson, "The Iron Ring"; D. Allan Bromley to the author, note dated March 24, 1995.

16. Jamieson, "The Iron Ring"; Wikipedia, "Iron Ring"; Osman, "The Iron Ring"; G. J. Thomson to author, fax dated March 16, 1995. A tradition of engineers wearing a ring made of iron on a gold base was established in the late nineteenth century at the Swedish Royal Institute of Technology. In 1927, a group of mechanical engineering students from the German Technical University in Prague visited Sweden, where they learned of the Swedish custom. Upon returning to Prague, they instituted their own ring tradition. Ernst R. G. Eckert to author, letter dated May 24, 1995; Wikipedia, "The Ritual of the Calling of an Engineer," http://en.wikipedia.org/wiki/The_Ritual_of_the_Calling_of_an_Engineer.

17. Wikipedia, "Iron Ring."

18. Emanuel D. Rudolph, "Obituaries of Members of the Ohio Academy of Science: Report of the Necrology Committee," [1991], s.v. Lloyd Adair Chacey (1899–1990), https://kb.osu.edu/dspace/bitstream/1811/23480/1/V091N5_221.pdf; Oscar T. Lyon, Jr., "Nothing New," letter to the editor, *ASCE News*, Oct. 1989.

19. Lyon, "Nothing New"; Homer T. Borton, "The Order of the Engineer," *The Bent of Tau Beta Pi*, Spring 1978, pp. 35–37.

20. Borton, "The Order of the Engineer," p. 35; compare "The Order of the Engineer," http://www.order-of-the-engineer.org; G. N. Martin to Lloyd A. Chacey, letter dated July 17, 1972, reproduced in *Manual for Engineering Ring Presentations*, rev. ed. (Cleveland: Order of the Engineer, 1982), p. 6.

21. http://www.order-of-the-engineer.org. See also *Manual for Engineering Ring Presentations*, p. 2. Because I have chronic arthritis, which has on occasion caused my fingers and their joints to swell, I do not wear rings of any kind. For this reason, I have not recited the Obligation of an Engineer, nor do I plan to do so.

22. Order of the Engineer, *Order of the Engineer*, booklet (Cleveland, Ohio: Order of the Engineer, 1981).

23. Order of the Engineer, *Manual for Engineers Ring Presentations*, pp. 18–20.

24. Order of the Engineer, "About the Order," http://www.order-of-the-engineer.org.

25. Paul H. Wright, *Introduction to Engineering*, 2nd ed. (New York: Wiley, 1994), Fig. 3.1; Connie Parenteau and Glen Sutton, "The Ritual of the Calling of an Engineer (Iron Ring Ceremony)," PowerPoint presentation, Eng G 400, University of Alberta, 2006, http://www.engineering.ualberta.ca/pdfs/IronRing.pdf.

26. Parenteau and Sutton, "The Ritual of the Calling of an Engineer."

27. Ibid.; Rudyard Kipling, "Hymn of Breaking Strain," *The Engineer*, March 15, 1935; see also Rudyard Kipling, *Hymn of the Breaking Strain* (Garden City, N.Y.: Doran, 1935); http://etext.lib.virginia.edu/etcbin/toccer-new2?id=KipBrea.sgm&images=images/modeng&data=/texts/english/modeng/parsed&tag=public&part=all.

28. Parenteau and Sutton, "The Ritual of the Calling of an Engineer."

29. Wikipedia, "Iron Ring."

30. Order of the Engineer, *The Order of the Engineer*, p. 9; M. G. Britton, ". . . And Learning from Failure," *The Keystone Professional* (Association of Professional Engineers and Geoscientists of the Province of Manitoba), Spring 2010, p. 7. The ring ceremony I attended took place at the 2009 Forensic Engineering Congress, whose proceedings are contained in *Forensic Engineering 2009: Pathology of the Built Environment*, ed. Shen-en Chen et al. (Reston, Va.: American Society of Civil Engineers, 2009).

31. Although the tradition of wearing an Iron Ring is still most commonly associated with Canadian engineers, Scandinavian and other European engineers have followed similar traditions. Borton, "The Order of the Engineer," p. 36; *The Order of the Engineer*, p. 7; Carol Reese to author, email message, June 30, 2010;

Glen Sutton to author, email message, Feb. 1, 2010; Yngve Sundström to author, May 25, 1995; R. G. Eckert to author, May 24, 1995.

9. Before, during, and after the Fall

1. See, e.g., "Tacoma Narrows Bridge Collapse Gallopin' Gertie," http://www.youtube.com/watch?v=j-zczJXSxnw.

2. "Big Tacoma Bridge Crashes 190 Feet into Puget Sound," *New York Times,* Nov. 8, 1940, p. 1.

3. Richard Scott, *In the Wake of Tacoma: Suspension Bridges and the Quest for Aerodynamic Stability* (Reston, Va.: ASCE Press, 2001), p. 41.

4. Richard S. Hobbs, *Catastrophe to Triumph: Bridges of the Tacoma Narrows* (Pullman: Washington State University Press, 2006), pp. 9–11; Scott, *In the Wake of Tacoma,* p. 41.

5. "Calling the Role of Key Construction Men Captured at Guam," *Pacific Builder and Engineer,* Dec. 1945, p. 48; Clark H. Eldridge, "The Tacoma Narrows Suspension Bridge," *Pacific Builder and Engineer,* July 6, 1940, pp. 34–40.

6. John Steele Gordon, "Tacoma Narrows Bridge Is Falling Down," AmericanHeritage.com, Nov. 7, 2007, http://www.americanheritage.com/articles/web/20071107-tacoma-narrows-bridge-leon-moisseiff-galloping-gertie.shtml; Clark H. Eldridge, "The Tacoma Narrows Bridge," *Civil Engineering,* May 1940, pp. 299–302.

7. Henry Petroski, *Engineers of Dreams: Great Bridge Builders and the Spanning of America* (New York: Knopf, 1995), pp. 297–300; Hobbs, *Catastrophe to Triumph,* p. 12.

8. Hobbs, *Catastrophe to Triumph,* pp. 17–20.

9. Ibid., pp. 58–60.

10. "Big Tacoma Bridge Crashes," p. 1.

11. Hobbs, *Catastrophe to Triumph,* pp. 64–65.

12. Washington State Department of Transportation, "Tacoma Narrows Bridge: Tubby Trivia," http://www.wsdot.wa.gov/tnbhistory/tubby.htm.

13. "Big Tacoma Bridge Crashes," pp. 1, 5; "Charges Economy on Tacoma Bridge," *New York Times,* Nov. 9, 1940, p. 19.

14. "Big Tacoma Bridge Crashes," p. 5; "Charges Economy on Tacoma Bridge"; "A Great Bridge Falls," *New York Times,* Nov. 9, 1940, p. 16.

15. Othmar H. Ammann, Theodore von Kármán, and Glenn B. Woodruff, "The Failure of the Tacoma Narrows Bridge," report to Federal Works Agency, March 28, 1941; Scott, *In the Wake of Tacoma,* pp. 53–55.

16. Delroy Alexander, "A Lesson Well Learnt," *Construction Today*, Nov. 1990, p. 46.

17. "Professors Spread the Truth about Gertie," *Civil Engineering*, Dec. 1990, pp. 19–20; K. Yusuf Billah and Robert H. Scanlan, "Resonance, Tacoma Narrows Bridge Failure, and Undergraduate Physics Textbooks," *American Journal of Physics* 59 (1991): 118–124.

18. Billah and Scanlan, "Resonance," p. 120.

19. Ibid., pp. 121–122. A more general technical treatment, including the coupling of vertical and torsional oscillations of bridge decks, is contained in Earl Dowell, ed., *A Modern Course in Aeroelasticity*, 2nd rev. and enlarged ed. (Dordrecht: Klewer Academic Publishers, 1989).

20. Billah and Scanlan, "Resonance," p. 123.

21. Hobbs, *Catastrophe to Triumph*, pp. 79–82, 127.

22. Washington State Department of Transportation, "Tacoma Narrows Bridge Connections," http://www.wsdot.wa.gov/tnbhistory/connections/connections4.htm.

23. Hobbs, *Catastrophe to Triumph*, p. 100; Lawrence A. Rubin, *Mighty Mac: The Official Picture History of the Mackinac Bridge* (Detroit: Wayne State University Press, 1958). For Steinman's analysis of the behavior of suspension-bridge decks in the wind, see David B. Steinman, "Suspension Bridges: The Aerodynamic Problem and Its Solution," *American Scientist*, July 1954, pp. 397–438, 460.

24. Hobbs, *Catastrophe to Triumph*, pp. 123–127.

25. Thomas Spoth, Ben Whisler, and Tim Moore, "Crossing the Narrows," *Civil Engineering*, Feb. 2008, pp. 38–47.

26. Spoth et al., "Crossing the Narrows," p. 40.

27. Ibid., pp. 43, 45.

28. Ibid., pp. 45, 47.

29. Tom Spoth, Joe Viola, Augusto Molina, and Seth Condell, "The New Tacoma Narrows Bridge—From Inception to Opening," *Structural Engineering International* 1 (2008): 26; Thomas G. Dolan, "The Opening of the New Tacoma Narrows Bridge: 19,000 Miles of Wire Rope," *Wire Rope News & Sling Technology*, Oct. 2007, p. 44.

30. Sheila Bacon, "A Tale of Two Bridges," *Constructor*, Sept.–Oct. 2007, http://constructor.agc.org/features/archives/0709–64.asp; see also Mike Lindblom, "High-Wire Act," *Seattle Times, Pacific Northwest Magazine*, Sept. 11, 2005, pp. 14–15; Spoth et al., "Crossing the Narrows," p. 43.

31. Lindblom, "High-Wire Act," p. 28.

32. "Tacoma Narrows Bridge History," *Seattle Times*, July 13, 2007, p. A14.

10. Legal Matters

1. For background and failure analysis of the I-35W bridge, see Barry B. Le-Patner, *Too Big to Fall: America's Failing Infrastructure and the Way Forward* (New York: Foster Publishing, 2010), pp. 3–26.

2. Henry Petroski, "The Paradox of Failure," *Los Angeles Times,* Aug. 4, 2007, p. A17.

3. Tudor Van Hampton, "Engineers Swarm on U.S. Bridges to Check for Flaws," *Engineering News-Record,* Aug. 20, 2007, p. 12; Aileen Cho, Tom Ichniowski, and William Angelo, "Engineers Await Tragedy's Inevitable Impacts," *Engineering News-Record,* Aug. 13, 2007, pp. 12–16; Ken Wyatt, "I-35W Bridge Was Overloaded," letter to the editor, *Civil Engineering,* June 2009, p. 8.

4. Tudor Van Hampton, "Federal Probe Eyes Gusset-Plate Design," *Engineering News-Record,* Aug. 20, 2007, pp. 10–11; Tom Ichniowski, "NTSB Cites Gussets and Loads in Collapse," *Engineering News-Record,* Nov. 24, 2008, pp. 60–61.

5. Christina Capecchi, "Work Starts on Minneapolis Bridge Replacement," *New York Times,* Nov. 2, 2007, p. A20; Michael C. Loulakis, "Appellate Court Validates I-35W Bridge Procurement," *Civil Engineering,* Oct. 2009, p. 88; Aileen Cho and Tudor Van Hampton, "Agency Awards Flatiron Team Twin Cities Replacement Job," *Engineering News-Record,* Oct. 15, 2007, p. 12; Tudor Van Hampton, "Minneapolis Bridge Rebuild Draws Fire," *Engineering News-Record,* Oct. 1, 2007, pp. 10–11.

6. Monica Davey and Mathew L. Wald, "Potential Flaw Found in Design of Fallen Bridge," *New York Times,* Aug. 9, 2007, p. A1.

7. Monica Davey, "Back to Politics as Usual, after Bridge Failure," *New York Times,* Aug. 16, 2007, pp. A1, A16.

8. Kevin L. Western, Alan R. Phipps, and Christopher J. Burgess, "The New Minneapolis I-35W Bridge," *Structure Magazine,* April 2009, pp. 32–34; Christina Capecchi, "Residents Divided on Design for New Span in Minneapolis," *New York Times,* Oct. 13, 2007, p. A8.

9. "Span of Control," Economist.com, May 20, 2009; Western et al., "New Minneapolis I-35W Bridge"; see also LePatner, *Too Big to Fall,* pp. 147–148.

10. Henry Fountain, "Concrete: The Remix," *New York Times,* March 31, 2009, pp. D1, D4; Western et al., "New Minneapolis I-35W Bridge."

11. Ichniowski, "NTSB Cites Gussets and Loads in Collapse."

12. "At I-35W, Engineers Develop Bridge-Collapse Scenario," *Engineering News-Record,* Nov. 5, 2007, p. 19.

13. Associated Press, "Minneapolis Bridge Victims Seek Punitive Damages,"

June 28, 2010. For another failure scenario, see LePatner, *Too Big to Fall*, pp. 206–207, note 32.

14. Mike Kaszuba, "One More I-35W Collapse Lawsuit to Come?" (Minneapolis) *Star Tribune*, July 5, 2009; Aileen Cho and Tudor Hampton, "I-35W Suit to Target Engineer, Contractor," *Engineering News-Record*, April 6, 2009, pp. 12–13.

15. Richard Korman, "Judge Declines to Dismiss Collapse Case against Jacobs," *Engineering News-Record*, Sept. 7, 2009, p. 12.

16. Gerald Sheine to author, e-mail messages dated Nov. 12 and 26, 2009.

17. Ibid.

18. "Firm Settles Suit over Minn. Bridge Collapse," *USA Today*, Aug. 24, 2010, p. 3A; Brian Bakst, "Firm to Pay $52.4M in Minneapolis Bridge Collapse," Associated Press, Aug. 24, 2010; "URS Agrees to Pay $52.4M to Settle Claims from Minn. Bridge Collapse," ENR.com, Aug. 23, 2010, http://enr.construction.com/yb/enr/article.aspx?story_id=148975524.

11. Back-Seat Designers

1. For some brief comments on whether the new millennium began with the year 2000 or 2001, see Henry Petroski, *Pushing the Limits: New Adventures in Engineering* (New York: Alfred A. Knopf, 2004), pp. 248–249.

2. On the Millennium Bridge, see, e.g., ibid., pp. 107–112.

3. Joseph Edward Shigley, *Machine Design* (New York: McGraw-Hill, 1956); Richard Gordon Budynas and J. Keith Nisbett, *Shigley's Mechanical Engineering Design*, 8th ed. (New York: McGraw-Hill, 2006); Richard G. Budynas and J. Keith Nisbett, *Shigley's Mechanical Engineering Design*, 9th ed. (New York: McGraw-Hill, 2011).

4. Nadine M. Post, "Third Exit Stair Could Make Highrises Too Costly to Build," *Engineering News-Record*, June 4, 2007, p. 13.

5. Joseph F. McCloskey, "Of Horseless Carriages, Flying Machines, and Operations Research," *Operations Research* 4 (1956) 2: 142. See also I. B. Holley, Jr., *The Highway Revolution, 1895–1925: How the United States Got out of the Mud* (Durham, N.C.: Carolina Academic Press, 2008).

6. John Lancaster, *Engineering Catastrophes: Causes and Effects of Major Accidents*, 2nd ed. (Boca Raton, Fla.: CRC Press, 2000), pp. 26–29.

7. Heather Timmons and Hari Kumar, "On India's Roads, a Grim Death Toll That Leads the World," *New York Times*, June 9, 2010, pp. A4, A8; Siddharth Philip, "One-Dollar Bribes for India Licenses Contribute to World's Deadliest Roads," Bloomberg.com, Nov. 30, 2010.

8. Ralph Nader, *Unsafe at Any Speed: The Designed-In Dangers of the American Automobile* (New York: Pocket Books, 1966), p. v; The Public Purpose, "Annual US Street & Highway Fatalities from 1957," http://www.publicpurpose.com/hwy-fatal57+.htm.

9. Nader, *Unsafe at Any Speed,* pp. 140–142.

10. Ibid., chap. 3, p. 74.

11. Ibid., pp. 225–230.

12. Quoted in ibid., pp. 129, 136, 152.

13. Committee on Trauma and Committee on Shock, "Accidental Death and Disability: The Neglected Disease of Modern Society" (Washington, D.C.: National Academy of Sciences and National Research Council, 1966), pp. 1, 5; see http://www.nap.edu/openbook.php?record_id=9978&page=5; Nader, *Unsafe at Any Speed,* p. 249.

14. Nader, *Unsafe at Any Speed,* pp. 199–200.

15. "President Johnson Signs the National Traffic and Motor Vehicle Safety Act," *This Day in History,* Sept. 9, 1966, http://www.history.com/this-day-in-history/president-johnson-signs-the-national-traffic-and-motor-vehicle-safety-act; "National Traffic and Motor Vehicle Saftey Act of 1966," enotes.com, http://www.enotes.com/major-acts-congress/national-traffic-motor-vehicle-safety-act.

16. "New NHTSA Study Finds U.S. Highway Deaths Lowest Since 1954," *Kelly Blue Book,* Mar. 12, 2010, http://www.kbb.com/car-news/all-the-latest/new-nhtsa-study-finds-us-highway-deaths-lowest-since-1954; "Motor Vehicle Traffic Fatalities & Fatality Rate: 1899–2003," http://www.saferoads.org/federal/2004/TrafficFatalities1899–2003; Michael Cooper, "Happy Motoring: Traffic Deaths at 61-Year Low," *New York Times,* April 1, 2011, p. A15.

17. Jerry L. Mashaw and David L. Harfst, "Regulation and Legal Culture: The Case of Motor Vehicle Safety," *Yale Journal on Regulation* 4 (1987): 257–258.

18. Ibid., pp. 262–264, 266–267; "National Traffic and Motor Vehicle Safety Act of 1966."

19. Mashaw and Harfst, "Regulation and Legal Culture," p. 276.

20. Jo Craven McGinty, "Poking Holes in Air Bags," *New York Times,* May 15, 2010, pp. B1, B4; Steven Reinberg, "Six Out of 7 Drivers Use Seat Belts: CDC," *Bloomberg Businessweek,* Jan. 4, 2011, http://www.businessweek.com/lifestyle/content/healthday/648501.html.

21. McGinty, "Poking Holes in Air Bags."

22. Nick Bunkley, "Toyota Concedes 2 Flaws Caused Loss of Control," *New York Times,* July 15, 2010, pp. B1, B4; Matthias Gross, *Ignorance and Surprise: Science, Society, and Ecological Design* (Cambridge, Mass.: MIT Press, 2010), p. 15;

Kimberly Kindy, "Vehicle Safety Bills Reflect Compromise between U.S. Legislators and Automakers," *Washington Post,* June 8, 2010, p. A15; "A Tougher Car Safety Agency," editorial, *New York Times,* July 31, 2010, p. A18.

23. Peter Whoriskey, "U.S. Report Finds No Electronic Flaws in Toyotas That Would Cause Acceleration," *Washington Post,* Feb. 9, 2011, http://www.washingtonpost.com/wp-dyn/content/article/2011/02/08/AR2011020800540_pf.html; Jayne O'Donnell, "Engineers Who Wrote Report Can't 'Vindicate' Toyota," *USA Today,* Feb. 8, 2011, http://www.usatoday.com/money/autos/2011-02-09-toyota09_VA1_N.htm.

24. San Francisco Bicycle Coalition, "Bridge the Gap!" http://www.sfbike.org/?baybridge.

25. Neal Bascomb, *The New Cool: A Visionary Teacher, His F.I.R.S.T. Robotics Team, and the Ultimate Battle of Smarts* (New York: Crown, 2010), pp. 39, 57.

26. Ibid., p. 216.

27. The Franklin Institute, "Edison's Lightbulb," http://www.fi.edu/learn/scitech/edison-lightbulb/edison-lightbulb.php?cts=electricity; Phrase Finder, "If at first you don't succeed. . . ," http://www.phrases.org.uk/bulletin_board/5/messages/266.html; Thomas H. Palmer, *Teacher's Manual* (1840), quoted in Bartlett's, pp. 393–394; see also Gregory Y. Titelman, *Random House Dictionary of Popular Proverbs and Sayings* (New York: Random House, 1996), p. 154.

28. Samuel Beckett, *Worstward Ho* (London: John Calder, 1983), p. 7. This quotation was called to my attention in William Grimson to the author, e-mail message dated Aug. 4, 2011.

12. Houston, *You* Have a Problem

1. R. P. Feynman, "Personal Observations on the Reliability of the Shuttle," at http://www.ralentz.com/old/space/feynman-report.html; Barry B. LePatner, *Too Big to Fall: America's Failing Infrastructure and the Way Forward* (New York: Foster Publishing, 2010), p. 89.

2. "Feynman O-Ring Junta Challenger," video clip, http://www.youtube.com/watch?v=KYCgotDV10c; quoted in Leonard C. Bruno, "Challenger Explosion," in *When Technology Fails: Significant Technological Disasters, Accidents, and Failures of the Twentieth Century,* ed. Neil Schlager (Detroit: Gale Research, 1994), p. 614.

3. Quoted in Bruno, "Challenger Explosion," p. 614.

4. "List of Space Shuttle Missions," http://en.wikipedia.org/wiki/List_of_space_shuttle_missions; Center for Chemical Process Safety, "Lessons from the Columbia Disaster," slide presentation, 2005, http://www.aiche.org/uploaded-

Files/CCPS/Resources/KnowledgeBase/Presentation_Rev_newv4.ppt#1079; Columbia Accident Investigation Board, *Report*, vol. 1, Aug. 2003, *http://caib.nasa. gov/news/report/volume1/default.html*; John Schwartz, "Minority Report Faults NASA as Compromising Safety," *New York Times*, Aug. 18, 2005, p. A18.

5. "Day 42: The Latest on the Oil Spill," *New York Times*, June 2, 2010, p. A16.

6. Brian Stelter, "Cooper Becomes Loud Voice for Gulf Residents," *New York Times*, June 18, 2010, p. A19; "The Gulf Oil-Spill Disaster Is Engineering's Shame," *Engineering News-Record*, June 7, 2010, p. 56.

7. Letters to the editor from Ronald A. Corso, Harry T. Hall, and William Livingston, *Engineering News-Record*, June 28, 2010, pp. 4–5.

8. For a book that focuses on management errors as a cause of failures, see James R. Chiles, *Inviting Disaster: Lessons from the Edge of Technology—An Inside Look at Catastrophes and Why They Happen* (New York: HarperCollins, 2001).

9. Transocean, "Deepwater Horizon: Fleet Specifications," http://www.deepwater.com/fw/main/Deepwater-Horizon-56C17.html?LayoutID=17; Transocean, "A Next Generation Driller Is Innovative," http://www.deepwater.com/fw/main/Home-1.html; Ian Urbina and Justin Gillis, "'We All Were Sure We Were Going to Die,'" *New York Times*, May 8, 2010, pp. A1, A13; Reuters, "Timeline—Gulf of Mexico Oil Spill," Reuters.com, June 3, 2010, http://www.reuters.com/article/idUSN0322326220100603. For a good retelling of events leading up to, during, and in the aftermath of the crisis on the *Deepwater Horizon*, see Joel Achenbach, *A Hole at the Bottom of the Sea: The Race to Kill the BP Oil Gusher* (New York: Simon & Schuster, 2011).

10. Reuters, "Timeline."

11. Pam Radtke Russell, "Crude Awakening," *Engineering News-Record*, May 10, 2010, pp. 10–11; Reuters, "Timeline"; Sam Dolnick and Henry Fountain, "Unable to Stanch Oil, BP Will Try to Gather It," *New York Times*, May 6, 2010, p. A20; H. Josef Hebert and Frederic J. Frommer, "What Went Wrong at Oil Rig? A Lot, Probers Find," (Durham, N.C.) *Herald-Sun*, May 13, 2010, pp. A1, A5.

12. Henry Fountain, "Throwing Everything, Hoping Some Sticks," *New York Times*, May 15, 2010, p. A12.

13. Dolnick and Fountain, "Unable to Stanch Oil"; Elizabeth Weise, "Well to Relieve Oil Leak Closes in on Target," *USA Today*, July 1, 2010, p. 4A; "Talk About a Mess," *New York Times*, May 23, 2020, Week in Review, p. 2; Reuters, "Timeline"; Shaila Dewan, "In First Success, a Tube Captures Some Leaking Oil," *New York Times*, May 17, 2010, pp. A1, A15.

14. Mark Long and Susan Daker, "BP Optimistic on New Oil Cap," *Wall Street Journal*, July 11, 2010, http://online.wsj.com/article/SB10001424052748703854045

75358893150368072.html; Richard Fausset, "BP Says It's Closer to Oil Containment," *Los Angeles Times*, July 12, 2010, http://articles.latimes.com/2010/jul/12/nation/la-na-0712-oil-spill-20100712; Tom Breen and Harry R. Weber, "BP Testing Delayed on Gulf Oil Fix," ENR.com, July 14, 2010, http://enr.construction.com/yb/enr/article.aspx?story_id=147380069; Richard Fausset and Nicole Santa Cruz, "BP's Test of Newly Installed Cap Is Put Off," *Los Angeles Times*, July 13, 2010, http://www.latimes.com/news/nationworld/nation/la-na-oil-spill-20100714,0,1234918.story.

15. Colleen Long and Harry R. Weber, "BP, Feds Clash over Reopening Capped Gulf Oil Well," Associated Press, July 18, 2010, http://news.yahoo.com/s/ap/20100718/ap_on_bi_ge/us_gulf_oil_spill;_ylt=AgzDbLVOelmfPXiuLl9voPCsoNUE;_ylu=X3oDMTNocXRkb2lwBGFzc2VoA2FwLzIwMTAwNzE4L3VzX2d1bGZfb2lsX3NwaWxsBGNjb2RlA21vc3Rwb3B1bGFyBGNwb3MDMDMgRwb3MDMDFX2Nva2U2UEc2VjA3luX3RvcF9zdG9yeQRzbGsDc2NpZW50aXN0oaXNoc2dl Noc2dl Noc2dl; Henry Fountain, "Cap Connector Is Installed on BP Well," *New York Times*, July 12, 2010, p. A11; Henry Fountain, "Critical Test Near for BP's New Cap," *New York Times*, July 13, 2010, p. A15; Henry Fountain, "In Revised Plan, BP Hopes to Keep Gulf Well Closed," *New York Times*, July 19, 2010, pp. A1, A12; Campbell Robertson and Henry Fountain, "BP Caps Its Leaking Well, Stopping the Oil after 86 Days," *New York Times*, pp. A1, A18.

16. Campbell Robertson and Clifford Krauss, "Gulf Spill Is the Largest of Its Kind, Scientists Say," *New York Times*, Aug. 3, 2010, p. A14; Clifford Krauss, "'Static Kill' of the Well Is Working, Officials Say," *New York Times*, Aug. 5, 2010, p. A17; Clifford Krauss, "With Little Fanfare, Well Is Plugged with Cement," *New York Times*, Aug. 6, 2010, p. A13; Michael Cooper, "Coverage Turns, Cautiously, to Spill Impact," *New York Times*, Aug 7, 2010, p. A8.

17. Justin Gillis, "U.S. Report Says Oil That Remains Is Scant New Risk," *New York Times*, Aug. 4, 2010, pp. A1, A14; William J. Broad, "Oil Spill Cleanup Workers Include Many Very, Very Small Ones," *New York Times*, Aug. 5, 2010, p. A17; Campbell Robertson, "In Gulf, Good News Is Taken with Grain of Salt," *New York Times*, Aug. 5, 2010, pp. A1, A17; Clay Dillow, "Gulf Oil Disaster Update: Up to 80% of the Crude May Still Be Lurking in the Water," *Popular Science*, Aug. 17, 2010, http://www.popsci.com/science/article/2010-08/gulf-oil-update-80-oil-may-still-be-lurking-water.

18. Weise, "Well to Relieve Oil Leak Closes in on Target"; Henry Fountain, "Hitting a Tiny Bull's-Eye Miles under the Gulf," *New York Times*, July 6, 2010, pp. D1, D4; "Relief Well Nears Point of Intercept," *New York Times*, Aug. 10, 2010, p. A15; Henry Fountain, "Relief Well to Proceed to Ensure Spill Is Over," *New York Times*, Aug. 14, 2010, A11.

19. Brian Winter, "Ideas Pour in to Try to Help BP Handle Gulf Oil Spill," USA Today.com, June 9, 2010; Adrian Cho, "One Ballsy Proposal to Stop the Leak," Sciencemag.org, June 16, 2010, http://news.sciencemag.org/scienceinsider/2010/06/one-ballsy-proposal-to-stop-the-html; Christopher Brownfield, "Blow Up the Well to Save the Gulf," New York Times, June 22, 2010, p. A27; William J. Broad, "Nuclear Option on Oil Spill? No Way, U.S. Says," New York Times, June 3, 2010, pp. A1, A22.

20. Henry Fountain, "Far from the Ocean Floor, the Cleanup Starts Here," New York Times, May 18, 2010, p. D4; Joel Achenbach and Steven Mufson, "Engineers Trying Multiple Tactics in Battle to Plug Oil Well in Gulf of Mexico," Washington Post, May 11, 2010, p. A04; Richard Simon and Jill Leovy, "BP to Try Smaller Dome against Oil Leak," Los Angeles Times, May 11, 2010, http://www.latimes.com/news/nationworld/nation/la-na-oil-spill-20100511,0,645089.story; Clifford Krauss and Jackie Calmes, "Little Headway Is Made in Gulf as BP Struggles to Halt Oil Leak," New York Times, May 29, 2010, pp. A1, A13; "Government to Run Response Web Site," New York Times, July 5, 2010, p. A10; see also, John M. Broder, "Energy Secretary Emerges to Take a Commanding Role in Effort to Corral Well," New York Times, July 17, 2010, p. A11.

21. David Barstow et al., "Regulators Failed to Address Risks in Oil Rig Fail-Safe Device," New York Times, June 20, 2010; Henry Fountain, "BP Discussing a Backup Strategy to Plug Well," New York Times, June 29, 2010, p. A20.

22. David Barstow et al., "Between Blast and Spill, One Last, Flawed Hope," New York Times, June 21, 2010, pp. A1, A18–A20.

23. Barstow et al., "Between Blast and Spill," p. A18; Steven J. Coates to author, email dated June 8, 2010; Robbie Brown, "Another Rig's Close Call Altered Rules, Papers Say," New York Times, Aug. 17, 2010, p. A19.

24. Ian Urbina, "Oil Rig's Owner Had Safety Issue at 3 Other Wells," New York Times, Aug. 5, 2010, pp. A1, A16.

25. Pam Radtke Russell, "Investigations Expand List of BP's Drill-Program Failures," Engineering News-Record, June 28, 2010, p. 13.

26. Robbie Brown, "Siren on Oil Rig Was Kept Silent, Technician Says," New York Times, July 24, 2010, pp. A1, A11.

27. Russell, "Investigations Expand List of BP's Drill-Program Failures"; Reuters, "Timeline"; Barstow et al., "Between Blast and Spill," p. A19; Kevin Spear, "Did BP Make the Riskier Choice?" ENR.com, May 23, 2010, http:///enr.construction.com/yb/enr/article.aspx?story_id=145300960; Jennifer A. Dlouhy, "Spill Report: It Could Happen Again," Houston Chronicle, Jan. 5, 2011, http://www.chron.com/disp/story.mpl/business/7367856.html.

28. Campbell Robertson, "Efforts to Repel Gulf Oil Spill Are Described as

Chaotic," *New York Times,* June 15, 2010, pp. A1, A16; Jim Tankersley, Raja Abdulrahim, and Richard Fausset, "BP Makes Headway in Containing Oil Leak," *Los Angeles Times,* May 17, 2010, http://www.latimes.com/news/nationworld/la-na-oil-spill-20100517,0,1038311.story; Barstow et al., "Between Blast and Spill," p. A20; Juliet Eilperin, "U.S. Oil Drilling Agency Ignored Risk Warnings," *Washington Post,* May 25, 2010, pp. A1, A4; Kevin Giles, "St. Croix Bridge Plan Evaluation Slogged Down by Gulf Oil Leak" (Minneapolis) *Star Tribune,* June 27, 2010, Star-Tribune.com.

29. Hebert and Frommer, "What Went Wrong at Oil Rig?" pp. A1, A5; Robertson and Krauss, "Gulf Spill Is the Largest of Its Kind"; Justin Gillis, "Doubts Are Raised on Accuracy of Government's Spill Estimate," *New York Times,* May 14, 2010, pp. A1, A13; Tom Zeller, Jr., "Federal Officials Say They Vastly Underestimated Rate of Oil Flow into Gulf," *New York Times,* May 28, 2010, p. A15; Clifford Krauss and John M. Broder, "After a Setback, BP Resumes Push to Plug Oil Well," *New York Times,* May 28, 2010, pp. A1, A14; Justin Gillis and Henry Fountain, "Rate of Oil Leak, Still Not Clear, Puts Doubt on BP," *New York Times,* June 8, 2010, pp. A1, A18; Justin Gillis and Henry Fountain, "Experts Double Estimated Rate of Spill in Gulf," *New York Times,* June 11, 2010, pp. A1, A19; Joel Achenbach and David Fahrenthold, "Oil-Spill Flow Rate Estimate Surges to 35,000 to 60,000 Barrels a Day," *Washington Post,* June 15, 2010; John Collins Rudolf, "BP Is Planning to Challenge Federal Estimates of Oil Spill," *New York Times,* Dec. 4, 2010, p. A15.

30. "Historians Debate 'Worst Environmental Disaster' in U.S.," *Washington Post,* June 23, 2010; Justin Gillis, "Where Gulf Spill Might Place on the Roll of Great Disasters," *New York Times,* June 19, 2010, pp. A1, A10. A live video feed of oil emerging from the blowout preventer was available at http://www.bp.com/liveassets/bp_internet/globalbp/globalbp_uk_english/homepage/STAGING/local_assets/bp_homepage/html/rov_stream.html.

31. Campbell Robertson, "Gulf of Mexico Has Long Been Dumping Site," *New York Times,* July 30, 2010, pp. A1, A14; Robertson and Krauss, "Gulf Spill Is the Largest of Its Kind"; Urbina and Gillis, "'We All Were Sure We Were Going To Die.'"

32. United Press International, "Concrete Casing Flaws Eyed in Gulf Rig Explosion," *ENR.com,* May 23, 2010, http://enr.construction.com/yb/enr/article.aspx?story_id=145305139; Justin Gillis and John M. Broder, "Nitrogen-Cement Mix Is Focus of Gulf Inquiry," *New York Times,* May 11, 2010, p. A13; Ian Urbina, "BP Officials Took a Riskier Option for Well Casing," *New York Times,* May 27, 2010, pp. A1, A20; Matthew L. Wald, "Seeking Clues to Explosion, Experts Hope to Raise Rig's Remnants from Sea Floor," *New York Times,* June 9, 2010, p. A14.

33. Susan Saulny, "Tough Look Inward on Oil Rig Blast," *New York Times,*

May 12, 2010, p. A14; Ian Urbina, "U.S. Said to Allow Drilling Without Needed Permits," *New York Times,* May 14, 2010, pp. A1, A12.

34. John M. Broder, "U.S. to Split Up Agency Policing the Oil Industry," *New York Times,* May 12, 2010, pp. A1, A14.

35. National Academies, "Events Preceding Deepwater Horizon Explosion and Oil Spill Point to Failure to Account for Safety Risks and Potential Dangers," news release, Nov. 17, 2010.

36. Peter Baker, "Obama Gives a Bipartisan Commission Six Months to Revise Drilling Rules," *New York Times,* May 23, 2010, p. 16; Gerald Shields, "New Gulf Drilling Moratorium Issued," *The* (Baton Rouge, La.) *Advocate,* July 13, 2010, p. 1A; Russell Gold and Ben Casselman, "Far Offshore, a Rash of Close Calls," *Wall Street Journal,* Dec. 8, 2010, http://online.wsj.com/article/SB10001424052748 703989004575652714091006550.html?mod=WSJ_hp_MIDDLETopStories.

37. Dlouhy, "Spill Report."

38. Ben Casselman and Russell Gold, "Device's Design Flaw Let Oil Spill Freely," *Wall Street Journal,* March 24, 2011; Clifford Krauss and Henry Fountain, "Report on Oil Spill Pinpoints Failure of Blowout Preventer," *New York Times,* March 24, 2011, p. A18.

39. Final report and BP response quoted in John M. Broder, "Report Links Gulf Oil Spill to Shortcuts," *New York Times,* Sept. 15, 2011, p. A25; Pam Radtke Russell, "Final Deepwater Report Released," *Engineering News-Record,* Sept. 26, 2011, p. 12.

40. See Alistair Walker and Paul Sibly, "When Will an Oil Platform Fail?" *New Scientist,* Feb. 12, 1976, pp. 326–328; Gold and Casselman, "Far Offshore."

13. Without a Leg to Stand On

1. Vitruvius, *Ten Books on Architecture,* X, 2, 1–10; J. G. Landels, *Engineering in the Ancient World,* rev. ed. (Berkeley: University of California Press, 2000), pp. 84–85; for an excellent survey article, see Wikipedia, "Crane (machine)," http://en.wikipedia.org/wiki/Crane_(machine).

2. Georgius Agricola, *De re metallica,* trans. Herbert Clark Hoover and Lou Henry Hoover (New York: Dover Publications, 1950); Agostino Ramelli, *Diverse and Ingenious Machines of Agostino Ramelli,* trans. and ed. Martha Teach Gnudi and Eugene S. Ferguson (Baltimore: Johns Hopkins University Press, 1976); David de Haan, "The Iron Bridge—New Research in the Ironbridge Gorge," *Industrial Archaeology Review* 26 (2004), 1: 3–18; Nathan Rosenberg and Walter G. Vincenti, *The Britannia Bridge: The Generation and Diffusion of Technological Knowledge* (Cambridge, Mass.: MIT Press, 1978).

3. Tudor Van Hampton, "Feds Propose Crane Safety Rules, Operator Certifi-

cation Scheduled," ENR.com, Oct. 9, 2008, http://enr.construction.com/news/
safety/archives/081009.asp; Tudor Van Hampton, "Out of the Blind Zone," *Engineering News-Record,* Dec. 4, 2006, pp. 24–26; Tom Ichniowski, "Construction Deaths Down 16% in 2009, but Fatality Rates Stays Flat," ENR.com, Aug. 18, 2010, http://enr.ecnext.com/coms2/article_bmsh100819ConstDeathsD; see also Mohammad Ayub, "Structural Collapses during Construction," *Structure,* Dec. 2010, pp. 12–14.

4. Liz Alderman, "Real Estate Collapse Spells Havoc in Dubai," *New York Times,* Oct. 7, 2010, p. B7; Blair Kamin, "The Tallest Building Ever—Brought to You by Chicago; Burj Dubai's Lead Architect, Adrian Smith, Personifies City's Global Reach," *Chicago Tribune,* Jan. 2, 2010, http://featuresblogs.chicagotribune.com/theskyline/2010/01/the-tallest-building-everbrought-to-you-by-chicago-burj-dubais-lead-architect-adrian-smith-personifi.html.

5. See, e.g., Clifford W. Zink, *The Roebling Legacy* (Princeton, N.J.: Princeton Landmark Publications, 2011), p. 282.

6. Tudor Van Hampton, "Cranes Enabled Faster, Safer Construction in Tall Buildings," ENR.ecnext.com, Aug. 20, 2008.

7. My writing about tower cranes was prompted by email messages, one of which included a photo dated c. 1944 showing bomb damage to the port of Trieste. What today we call tower cranes are clearly visible above the damaged buildings. Hart Lidov to the author, email messages dated Nov. 23, 2002, and April 25, 2004.

8. Tudor Van Hampton, "Cranes and Cultures Clash in Dubai," *Engineering News-Record,* Dec. 4, 2006, p. 27.

9. "Records: World's Largest Tower Crane," *Engineering News-Record,* supplement, Dec. 2004, p. 16; "K-10000 Tower Crane Operating Speeds—U.S.," http://www.towercrane.com/K-10000_tower_cranes_24_00.htm; Tim Newcomb, "Massive Krøll Tower Crane Supports Seattle Tunnel Job," *Engineering News-Record,* Sept. 19, 2011, p 27.

10. Tudor Van Hampton et al., "Crane Anxiety Towers from Coast to Coast," *Engineering News-Record,* June 16, 2008, pp. 10–12; Richard Korman, "An Accident in Florida Shows a Break in the Decision Chain," *Engineering News-Record,* July 30, 2007, pp. 24–28.

11. Korman, "An Accident in Florida."

12. Ibid.

13. Ibid.; Van Hampton, "Out of the Blind Zone," p. 25.

14. Robert D. McFadden, "Crane Collapses on Manhattan's East Side, Killing 4," *New York Times,* March 16, 2008, p. A1.

15. William Neuman, "Failure of Nylon Strap Is Suspected in Crane Collapse," *New York Times*, March 18, 2008, p. C14; Tom Sawyer, "Crane-Accident Probe Targets Nylon Slings," *Engineering News-Record*, March 24, 2008, pp. 10–12.

16. Damien Cave, "Two Workers Are Killed in Miami Crane Accident," *New York Times*, March 26, 2008, p. A19.

17. James Barron, "Crane Collapse at New York Site Kills 2 Workers," *New York Times*, May 31, 2008, pp. A1, A16.

18. "Off the Hook," editorial, *Engineering News-Record*, June 9, 2008, p. 88; Eileen Schwartz, "Crane Failures Foul Up Texas' Already-Poor Safety Record," *Engineering News-Record*, June 23, 2008, pp. 98–99.

19. William Neuman and Anemona Hartocollis, "Inspector Is Charged with Filing False Report before Crane Collapse," *New York Times*, March 21, 2008, p. A20; William Neuman, "New York Tightens Regulation on Cranes," *New York Times*, March 26, 2008, p. B1; Diane Cardwell and Charles V. Bagli, "Building Dept. Head Resigns Her Post," *New York Times*, April 23, 2008, p. A22; Charles V. Bagli, "Amid Boom, a Battle over Buildings Chief's Qualifications," *New York Times*, June 11, 2008, pp. B1, B4; Sharon Otterman, "City Proposes More Regulations to Improve Construction Safety," *New York Times*, June 25, 2008, p. B3.

20. Dennis St. Germain, "Hidden Damage in Slings, Corrosion and Ultraviolet Light," *Wire Rope News & Sling Technology*, June 2010, pp. 12, 14, 16.

21. William K. Rashbaum, "Analysis of Crane Collapse Blames Improper Rigging," *New York Times*, March 12, 2009, p. A24; Nadine M. Post, "Climbing Crane Not Properly Secured, Says Manufacturer," *Engineering News-Record*, Oct. 9, 2006, p. 12; "Rigger Used Half the Hardware Than Crane's Maker Required," *Engineering News-Record*, March 23, 2009, p. 20; Jennifer Peltz, "Witness: New Straps Supplied before NYC Crane Fell," Associated Press story, July 1, 2010.

22. John Eligon, "Engineer Testifies Crane Rigger Is Careful," *New York Times*, July 10, 2010, p. A15; John Eligon, "Rigging Contractor Is Acquitted in the Collapse of a Crane," *New York Times*, July 23, 2010, p. A17.

23. Charles V. Bagli, "City Fined over Information on Fatal Crane Collapse," *New York Times*, April 7, 2010, p. A24; Tudor Van Hampton, "Crane-Failure Case Heading to Court," *Engineering News-Record*, March 15, 2010, pp. 10–11; Tudor Van Hampton, "What the Lomma Case Means to You," *Engineering News-Record*, March 15, 2010, p. 52.

24. John Eligon, "Former Chief Crane Inspector Admits Taking Bribes for Lies," *New York Times*, March 24, 2010, p A23; William K. Rashbaum, "City Issues Controversial New Rules Regulating Cranes at Construction Sites," *New York Times*, Sept. 20, 2008, p. B3; Tudor Van Hampton, "Proposed Crane Rule Gets

Mixed Reviews," *Engineering News-Record,* Oct. 20, 2008, p. 11; see also Tudor Van Hampton, "Federal Safety Regulators to Boost Tower-Crane Checks," *Engineering News-Record,* April 28, 2008, p. 12.

25. Tom Ichniowski, "Construction Industry Gets Ready to Implement Crane Safety Rule," ENR.com, Aug. 4, 2010, http://enr.ecnext.com/comsite5/bin/comsite5.pl?page=enr_document&item_id=0271-57773&format_id=XML.

26. Brad Fullmer et al., "Razor-Thin Margins as Contractors Fight for Stimulus Projects," *Engineering News-Record,* June 29, 2009, pp. 16–18; Nick Zieminski, "The U.S. Jobs Sector Hit Hardest by the Recession, Construction, May Not Reach Bottom until Sometime Next Year," Reuters, March 18, 2009, http://www.reuters.com/article/idUSTRE52H6M620090318; Paul Davidson, "Construction Unemployment Still on the Rise," *USA Today,* Feb. 26, 2010, http://www.usatoday.com/money/economy/employment/2010-02-25-construction25_ST_N.htm.

27. Sewell Chan, "Bernanke Says He Failed to See Financial Flaws," *New York Times,* Sept. 3, 2010, p. B3.

14. History and Failure

1. Sir Alfred Pugsley, R. J. Mainstone, and R. J. M. Sutherland, "The Relevance of History," *Structural Engineer* 52 (1974): 441–445; discussion, *Structural Engineer* 53 (1974): 387–398.

2. On the Dee Bridge, see James Sutherland, "Iron Railway Bridges," in Michael R. Bailey, ed., *Robert Stephenson—The Eminent Engineer* (Aldershot, Hants: Ashgate, 2003), pp. 302–335. On the Tacoma Narrows Bridge, see Richard Scott, *In the Wake of Tacoma: Suspension Bridges and the Quest for Aerodynamic Stability* (Reston, Va.: ASCE Press, 2001).

3. P. G. Sibly and A. C. Walker, "Structural Accidents and Their Causes," *Proceedings of the Institution of Civil Engineers* 62 (1977), Part 1: 191–208; Paul Sibly, "The Prediction of Structural Failures" (Ph.D. thesis, University of London, 1977).

4. Henry Petroski, "Predicting Disaster," *American Scientist,* March-April 1993, pp. 110–113; for failure scenarios for cable-stayed bridges, see Uwe Starossek, *Progressive Collapse of Structures* (London: Thomas Telford, 2009).

5. For further speculation on potentially vulnerable bridge types, see Henry Petroski, *Success Through Failure: The Paradox of Design* (Princeton, N.J.: Princeton University Press, 2006), pp. 172–174.

6. Spiro N. Pollalis and Caroline Otto, "The Golden Gate Bridge: The 50th Anniversary Celebration," Harvard University, Graduate School of Design, Laboratory for Construction Technology, Publication No. LCT-88–4, Nov. 1988, http://

www.goldengatebridge.org/research/documents/researchpaper_50th.pdf; Masayuki Nakao, "Closure of Millennium Bridge," *Failure Knowledge Database/100 Selected Cases,* http://shippai.jst.go.jp/en/Detail?fn=2&id=CA1000275.

7. See, e.g., Henry Petroski, "Design Competition," *American Scientist,* Nov.–Dec. 1997, pp. 511–515.

8. Deyan Sudjic et al., *Blade of Light: The Story of London's Millennium Bridge* (London: Penguin Press, 2001); Alexander N. Blekherman, "Swaying of Pedestrian Bridges," *Journal of Bridge Engineering* 10 (March-April 2005): 142; David McCullough, *The Great Bridge* (New York: Simon & Schuster, 1982), pp. 430–431; "Deadly Crush in Cambodia Tied to Bridge That Swayed," *New York Times,* Nov. 25, 2010, p. A22.

9. "World's Longest Stress Ribbon Bridge," *CE News,* June 2010, p. 15; Tony Sánchez, "Dramatic Bridge Provides a Natural Crossing," *Structural Engineering and Design,* June 2010, pp. 10–17, http://www.gostructural.com/magazine-article-gostructural_com-june-2010-dramatic_bridge_provides_a_natural_crossing-7918.html; Michael Stetz, "'One-of-a-Kind' Foot Bridge Still an Everyday Construction Site," (San Diego, Calif.) *Union-Tribune,* June 25, 2010, http://www.signonsandiego.com/news/2010/jun/25/one-of-a-kind-foot-bridge-still-an-everyday.

10. Sibly and Walker, "Structural Accidents and Their Causes."

11. Scott M. Adan and Ronald O. Hamburger, "Steel Special Moment Frames: A Historic Perspective," *Structure,* June 2010, pp. 13–14, http://www.structuremag.org/article.aspx?articleID=1079; Federal Emergency Management Agency, *World Trade Center Building Performance Study: Data Collection, Preliminary Observations, and Recommendations,* Report FEMA 403, May 2002, pp. 8–10.

12. Adan and Hamburger, "Steel Special Moment Frames."

13. Jim Hodges, "Generation to Generation: Filling the Knowledge Gaps," [NASA] *ASK Magazine,* 34 (Spring 2009): 6–9; Dave Lengyel, "Integrating Risk and Knowledge Management for the Exploration Systems Mission Directorate," ibid., pp. 10–12.

14. Justin Ray, "Taurus XL Rocket Launches Taiwan's New Orbiting Eye," SpaceflightNow.com, May 20, 2004, http://spaceflightnow.com/taurus/t7; NASA, "Taurus XL: Countdown 101," http://www.nasa.gov/mission_pages/launch/taurus_xl_count101.html; Rick Obenschain, "Anatomy of a Mishap Investigation," [NASA] *ASK Magazine,* 38 (Spring 2010): 5–8.

15. For more on why these bridge types are failure candidates to watch, see Henry Petroski, "Predicting Disaster," and Henry Petroski, *Success Through Failure: The Paradox of Design* (Princeton, N.J.: Princeton University Press, 2006), pp. 172–174.

16. Alistair Walker and Paul Sibly, "When Will an Oil Platform Fail?" *New Scientist*, Feb. 12, 1976, pp. 326–328.

17. Eugene S. Ferguson, *Engineering and the Mind's Eye* (Cambridge, Mass.: MIT Press, 1992). See also E. S. Ferguson, "The Mind's Eye: Nonverbal Thought in Technology," *Science* 197 (1977): 827–836.

18. See, e.g., John A. Roebling, *Final Report to the Presidents and Directors of the Niagara Falls Suspension Bridge and Niagara Falls International Bridge* (Rochester, N.Y.: Lee, Mann, 1855); O. H. Ammann, "George Washington Bridge: General Conception and Development of Design," *Transactions of the American Society of Civil Engineers* 97 (1933): 1–65; A. Pugsley, R. J. Mainstone, and R. J. M. Sutherland, "The Relevance of History," *Structural Engineer* 52 (1974): 441–445; see also discussion in *Structural Engineer* 53 (1975): 387–388.

19. Ralph Peck, "Where Has All the Judgement Gone?" Norges Geoteckiske Institutt, Publikasjon No. 134 (1981).

20. *Software Engineering Notes,* http://www.sigsoft.org/SEN; *The Risks Digest: Forum on Risks to the Public in Computers and Related Systems,* Peter G. Neumann, moderator, http://catless.ncl.ac.uk/Risks.

21. Jameson W. Doig and David P. Billington, "Ammann's First Bridge: A Study in Engineering, Politics, and Entrepreneurial Behavior," *Technology and Culture* 35 (1994), 3: 537–570; Vitruvius, *The Ten Books of Architecture,* trans. Morris Hicky Morgan (New York: Dover Publications, 1960), X, 16, 1–12; see also, e.g., *Transactions of the American Society of Civil Engineers* 97 (1933).

22. John A. Roebling, "Remarks on Suspension Bridges, and on the Comparative Merits of Cable and Chain Bridges," *American Railroad Journal, and Mechanics' Magazine* 6 (n.s.) (1841): 193–196.

23. Ibid.

24. Pauline Maier et al., *Inventing America: A History of the United States,* 2nd ed. (New York: Norton, 2006).

25. Sewell Chan, "Bernanke Says He Failed to See Financial Flaws," *New York Times,* Sept. 3, 2010, p. B3.

ILLUSTRATIONS

page 85   The massive testing machine was a centerpiece in Talbot Laboratory at the University of Illinois. (Photo courtesy of James W. Phillips.)

page 94   Computer-controlled table-top testing machines were convenient for studying the behavior of materials. (Photo courtesy of James W. Phillips.)

page 97   Donald E. Carlson was a professor of theoretical and applied mechanics at the University of Illinois and the author's dissertation advisor. (Photo courtesy of James W. Phillips.)

page 117   A plastic model of the cross-section of a Dee Bridge girder shows the so-called cavetto detailing where vertical web meets horizontal flange. (Photo courtesy of Peter R. Lewis.)

page 122   Tall smokestacks in Dundee, Scotland, survived the winds that blew on the December 1879 night that the high girders of the Tay Bridge fell into the river. (Digital image courtesy of Peter R. Lewis, made from contemporary print held in the Dundee City Library.)

page 142   A badly corroded Waldo-Hancock Bridge is dwarfed by the new Penobscot Narrows Bridge and Observatory. (Photo by Catherine Petroski.)

page 155   The Point Pleasant Bridge, known also as Silver Bridge, was of an unconventional design, in that some of the large steel links in its eyebar suspension chains also served as upper-chord members of the structure's stiffening trusses. (Postcard image courtesy of James E. Casto.)

page 176   The wreckage of the Quebec Bridge, which collapsed in August 1907 while still under construction, littered the banks of the St. Lawrence River. (From author's collection.)

page 179   D. Allan Bromley, Dean of Engineering at Yale University, proudly wore a ring that marked him as an engineer. (Photo courtesy of Yale University Office of Public Affairs.)

page 204   Washington State bridge engineer Clark H. Eldridge conferred with consultant Leon S. Moisseiff on the Tacoma Narrows Bridge. (From author's collection. Original photo courtesy of Shirley Eldridge; enhanced digital image courtesy of *American Scientist.*)

page 278   An explosion on the oil-drilling rig *Deepwater Horizon* set off a fire that led to the rig's sinking in about 5,000 feet of water in the Gulf of Mexico. (U.S. Coast Guard photo.)

page 312   A cluster of construction cranes served the Three Gorges Dam site on China's Yangtze River. (Photo courtesy of Kenneth L. Carper.)

INDEX

Italicized page numbers refer to figures and their captions.